MINI WEAPONS OF MASS DESTRUCTION™ 2

BUILD A SECRET AGENT ARSENAL

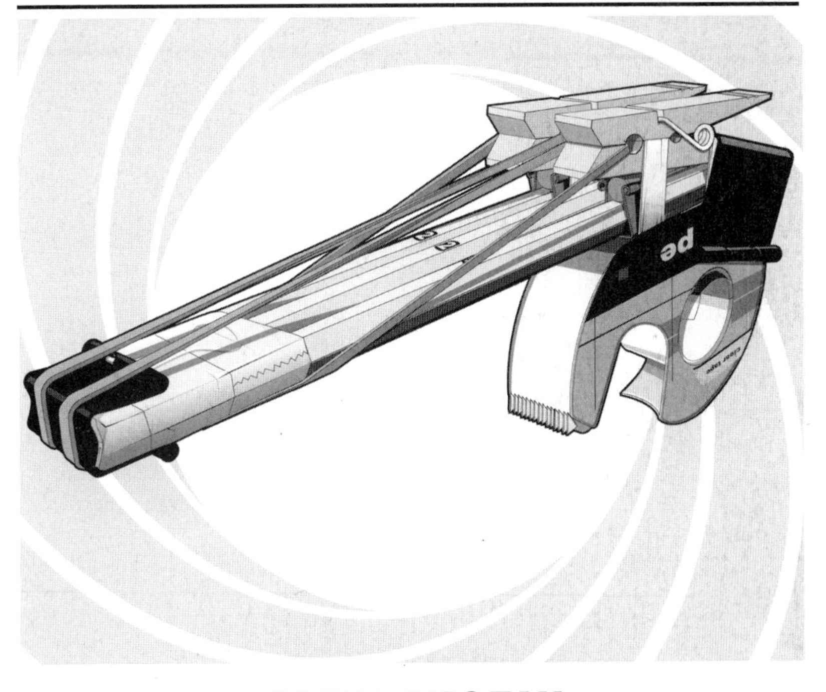

JOHN AUSTIN

Library of Congress Cataloging-in-Publication Data

Austin, John, 1978–

Miniweapons of mass destruction 2 : build a secret agent arsenal / John Austin.

p. cm.

Summary: "As a budding spy, what better way to conceal your clandestine activities than to miniaturize your secret agent arsenal? Hide a mini catapult in a breath mint tin. Turn a Tic Tac case into vest-pocket candy shooter. Or transform a milk jug cap onto a fake wristwatch that launches tiny paper darts. Toy designer and author John Austin provides detailed, step-by-step instructions with diagrams to show James Bondiacs how to build 35 different spy weapons and surveillance tools. All of the projects in MiniWeapons of Mass Destruction 2 are built from common household items—binder clips, playing cards, rubber bands, markers, clothespins, paper clips, and discarded packaging—clearly detailed on materials lists. In addition to movie-inspired sidearms and other "weapons," you'll find plans to construct periscopes, bionic ears, grappling hooks, and decipher pens. Once you have assembled your arsenal, the author provides a number of ideas on how to hide your stash—inside a deck of cards, a false-bottom soda bottle, or a cereal box briefcase—and targets for practicing your spycraft, including a flip-down firing range, a fake security camera, and sharks with laser beams. And if you think yourself more of an evil-genius in training, this book also has projects to keep you busy while you finish planning your volcano lair—a Q-pick blow gun, a paper throwing star, a bowler hat launcher, and more. Fluffy Persian cat not included"— Provided by publisher.

ISBN 978-1-56976-716-0 (pbk.)

1. Amusements. 2. Miniature weapons. 3. Handicraft. I. Title. II. Title: Build a secret agent arsenal. III. Title: Mini weapons of mass destruction 2. IV. Title: Miniweapons of mass destruction two.

GV1201.A8745 2011

790—dc23

2011027267

Cover and interior design: Jonathan Hahn

Illustrations: Austin Design, Inc.

Published by Chicago Review Press, Incorporated

814 North Franklin Street

Chicago, Illinois 60610

ISBN 978-1-56976-716-0

Printed in the United States of America

10 9 8 7 6 5

To those rookie special agents who have sworn an oath to protect and defend against all enemies foreign and domestic. So that you may bear the appropriate tools to obtain valuable confidential information for the greater good, maintain the secrecy of this agency and your own true identity, and conduct espionage operations with a minimum of risk, you have been granted the following guide to building a secret agent arsenal. Congratulations, agents, and may you be ready for anything!

Join the MiniWeapons army on Facebook:
MiniWeapons of Mass Destruction: Homemade Weapons Page

CONTENTS

INTRODUCTION

It's time to build a secret agent arsenal with *MiniWeapons of Mass Destruction 2*, the latest homemade weapons guide. You'll soon see the full potential of everyday items to be transformed into the gadgets, weaponry, and sophisticated tools every spy needs.

This collection of highly classified documents was gathered by our research and development department to provide field agents like you with quick countermeasures to any known threat. It is this division's job to equip our agents with mission-specific gear that will keep them from being captured—or worse.

Many of your missions—should you choose to accept them—are intense and rigorous, but acquiring the tools and weaponry to complete them shouldn't be. Each item in this book is built by following a bill of easy-to-locate materials and step-by-step instructions. And when you are not involved in top-secret spying or espionage, you can hone your marksmanship skills with the small library of simple targets in the final chapter. Always prepare for the unknown.

This is a book for spies of all ages. It pushes the laws of physics, inspires creativity, proposes experimentation, and fuels the imagination. Many of these projects are modeled after real-life gadgetry used by agents today but cost only pennies to assemble, giving you a *license to build!*

Keep in mind that this book is for entertainment purposes only. Please review the safety page for your personal protection. Build and use these projects at your own risk.

CONCEAL THIS BOOK!

According to our counterintelligence division, there have been reports that this book is a high-priority target and might be confiscated by unknown saboteurs (a.k.a. teachers). So for obvious reasons, you must make sure its highly classified contents remain out of reach of unauthorized hands. You can do this by quickly disguising your *MiniWeapons* book with a simple magazine cover.

First, find an outdated magazine you might normally read, then remove the binder staples and discard all the pages except for the cover.

Next, place the magazine cover facedown on the table, then fold the top and bottom of the magazine to align with the height of this book as shown in the illustration below. Make sure your creases are nice and crisp.

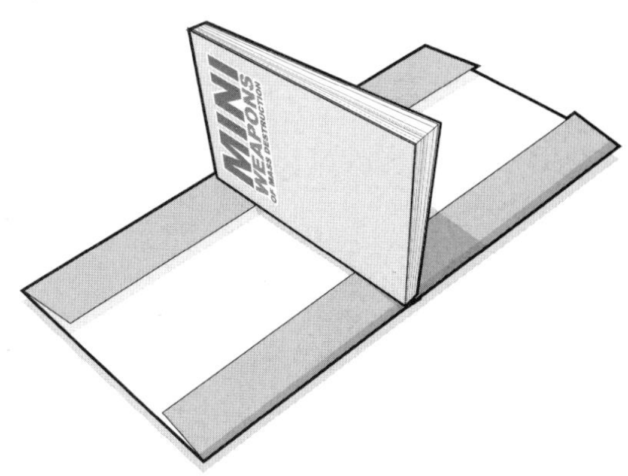

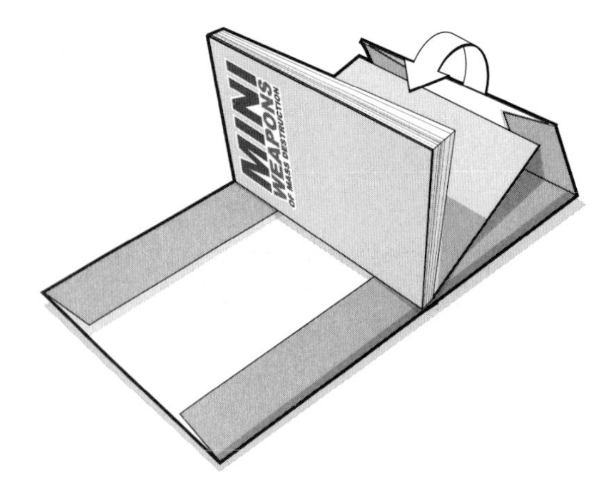

Now fold over the edges of the magazine cover above the *Mini-Weapons* book cover. Tape one side of the magazine cover together; do not tape your book. Repeat this step on the other side.

You have disguised your *MiniWeapons* book!

PLAY IT SAFE

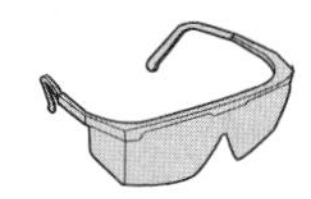

This book won't self-destruct, but homemade weapons might! Remember that the unexpected can happen. When building and firing Mini-Weapons, be responsible and take every safety precaution. Switching materials, substituting ammunition, assembling improperly, mishandling, targeting inaccurately, and misfiring could all cause harm. Like any good spy, you should always be prepared for the unknown. ***Eye protection is a must*** if you choose to experiment with any of these projects.

Most rookie spies don't have a sophisticated research and development (R&D) lab, so always be aware of your environment, including spectators and flammable materials, and be careful when handling the launchers. Arrows and darts have sharp points, and elastic and latex shooters fire projectiles with unbelievable force that can cause damage. Weapons–including handguns–should never be painted to look realistic; instead, choose a bright color–orange, red, or yellow. Ammo, no matter what the material, can cause harm. ***Never point a launcher at people, animals, or anything of value.*** And ***never*** take or transport any of these projects on public transportation, such as an airplane, bus, or train–these projects are to be used at home.

It is important to remember that since miniweaponry is homebuilt, it is not always accurate. Basic target blueprints and proposed printouts are available at the back of the book and downloadable at www.JohnAustinbooks.com; use these to test the accuracy of your Mini-Weapons, not random targets.

This book also contains plans for a Detonating Pen (page 179). Despite the name, this pen bomb could never be modified to do any serious damage. However, it is very loud. ***Ear protection is recommended for all "bombs."***

Some of the projects outlined in this book require various tools such as hobby knives, pocketknives, hot glue guns, and wire cutters that can cause injury if handled carelessly. Tools need your full attention–make safety your number-one priority. If you have trouble cutting, your knife may be dull or the selected material may be too hard; stop immediately and substitute one of the two. ***Junior agents should always be assisted by an adult when handling potentially harmful tools.***

Always be responsible when constructing and using miniweaponry. It is important that you understand that the author, the publisher, and the bookseller cannot and will not guarantee your safety. When you try the projects described here, you do so at your own risk. They are *not* toys!

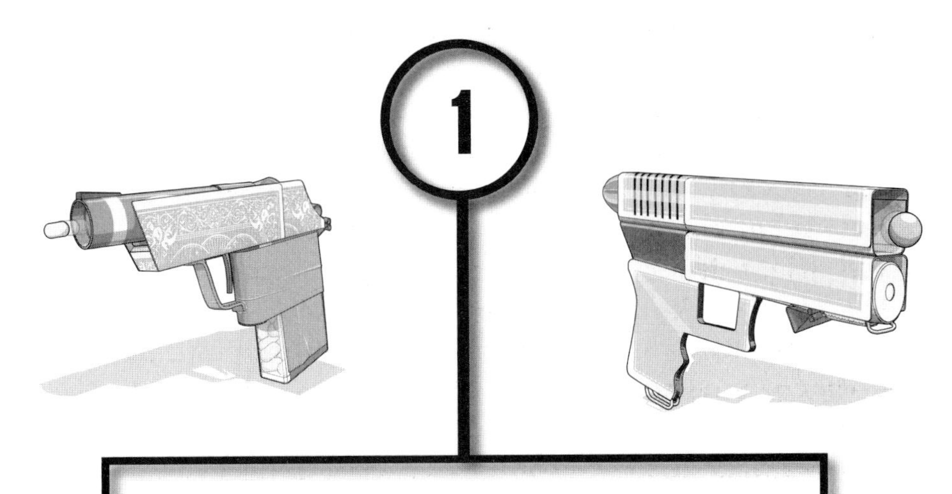

1 AGENT SIDEARMS

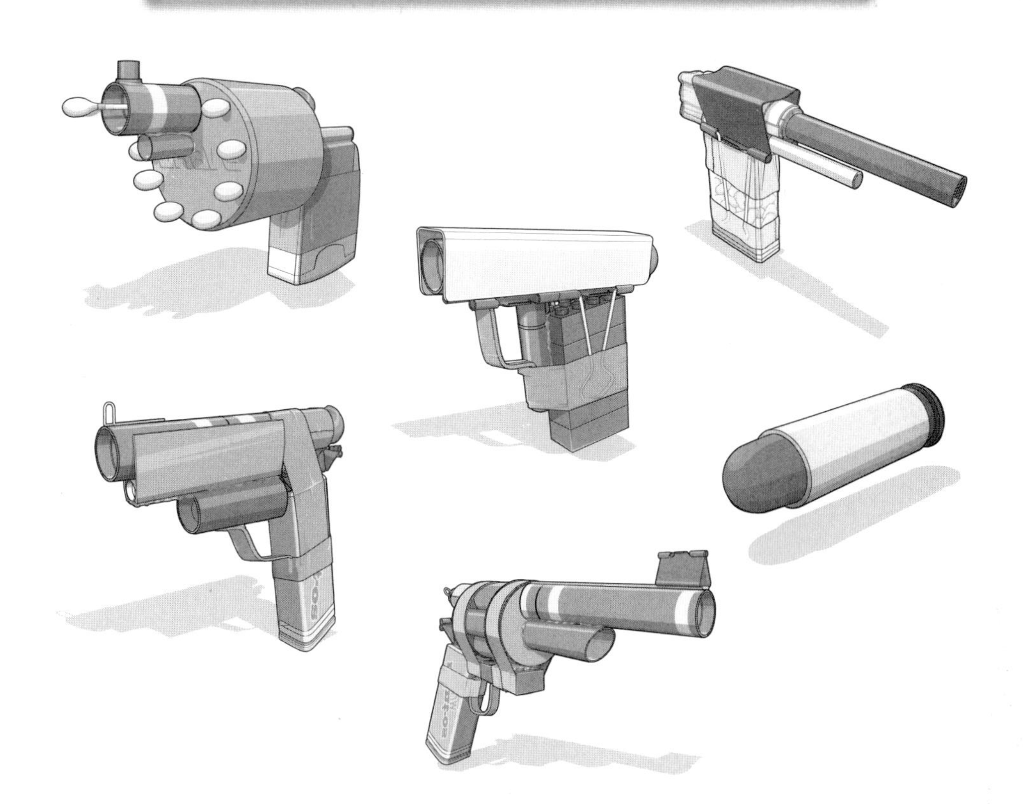

PPK TIC TAC

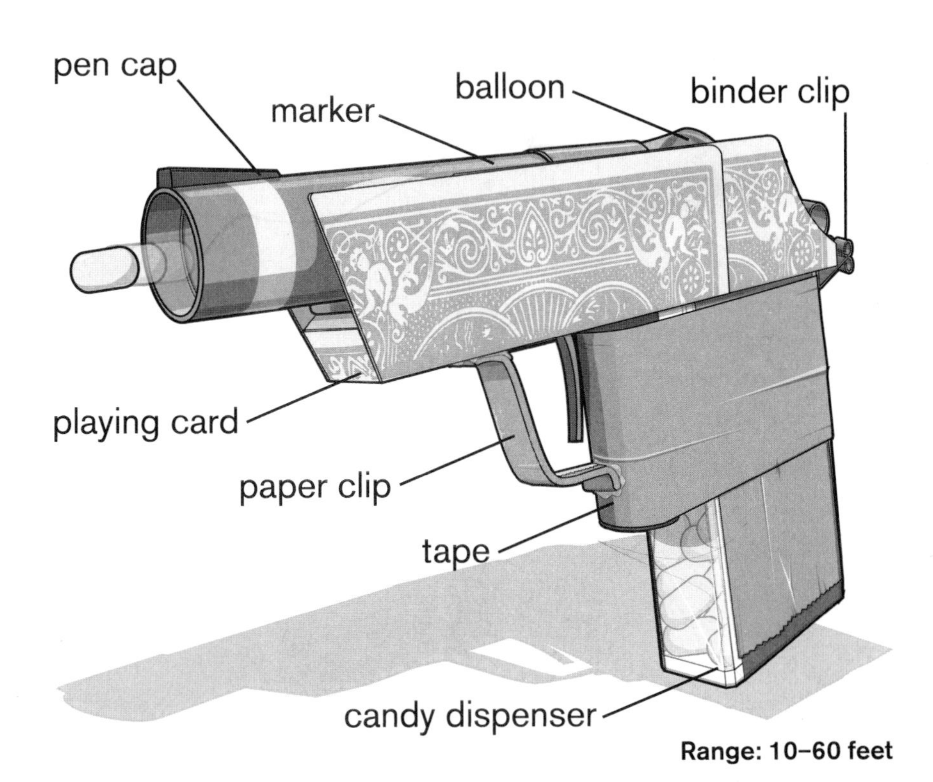

Range: 10–60 feet

Small and lightweight, the Walther PPK was the signature weapon of a very popular British secret agent. This homemade design features an incredible elastic firing range and an integrated candy dispenser, giving you an endless clip of minty bullets.

Supplies

1 standard art marker
1 small balloon
Duct tape
2 plastic pen caps
1 large paper clip
1 craft stick
1 Tic Tac container
1 small binder clip (19 mm)
2 playing cards

Tools

Safety glasses
Pliers
Hobby knife
Hot glue gun
Scissors

Ammo

1+ small, hard candies

Step 1

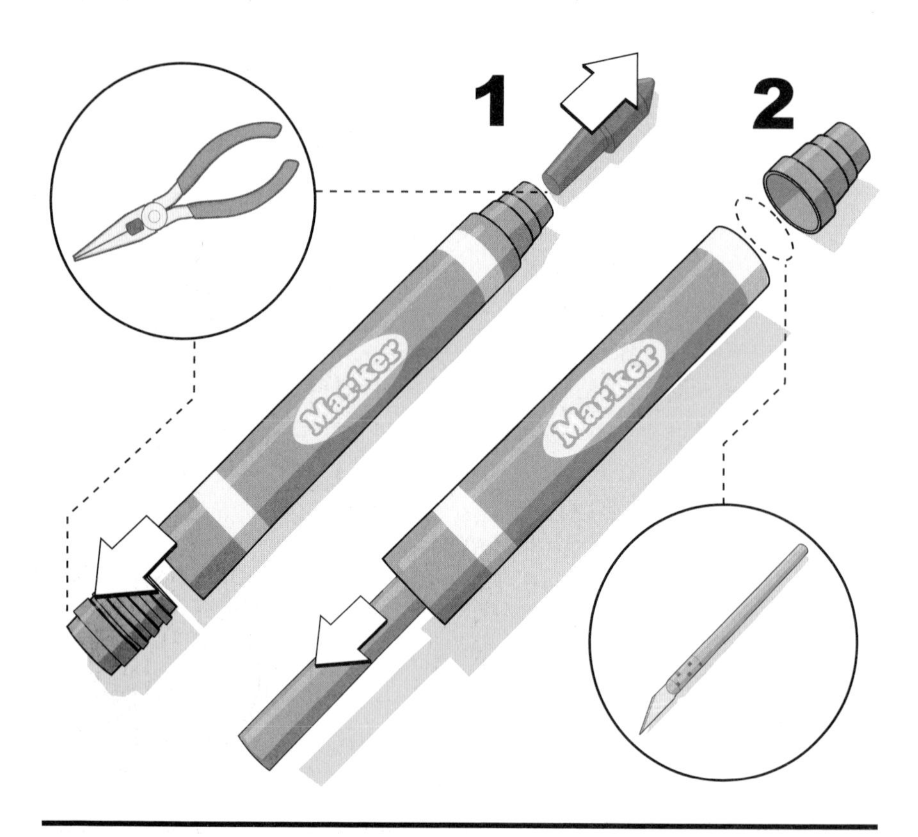

Locate a standard art marker; a recycled plastic marker is preferred because of the softness of the outer housing, but any marker will work. Using pliers, remove the marker's end cap by slightly twisting it from the housing. Then use the pliers to remove the marker nib–the writing point–by sliding it out the tip of the marker. Discard this nib to avoid a mess.

Using a hobby knife, pierce the marker housing below the plastic point as shown, and slowly rotate the marker. Rotating the housing, rather than the knife, will make cutting safer. Once this end has been removed, the internal ink cartridge should slide out. Discard this cartridge. (If the housing contains ink residue, rinse it out and dry it.)

The marker housing will become the barrel of your PPK Tic Tac. The housing pathway should be unobstructed by any plastic fragments that might have broken off during cutting. Save the marker cap for a later step.

Step 2

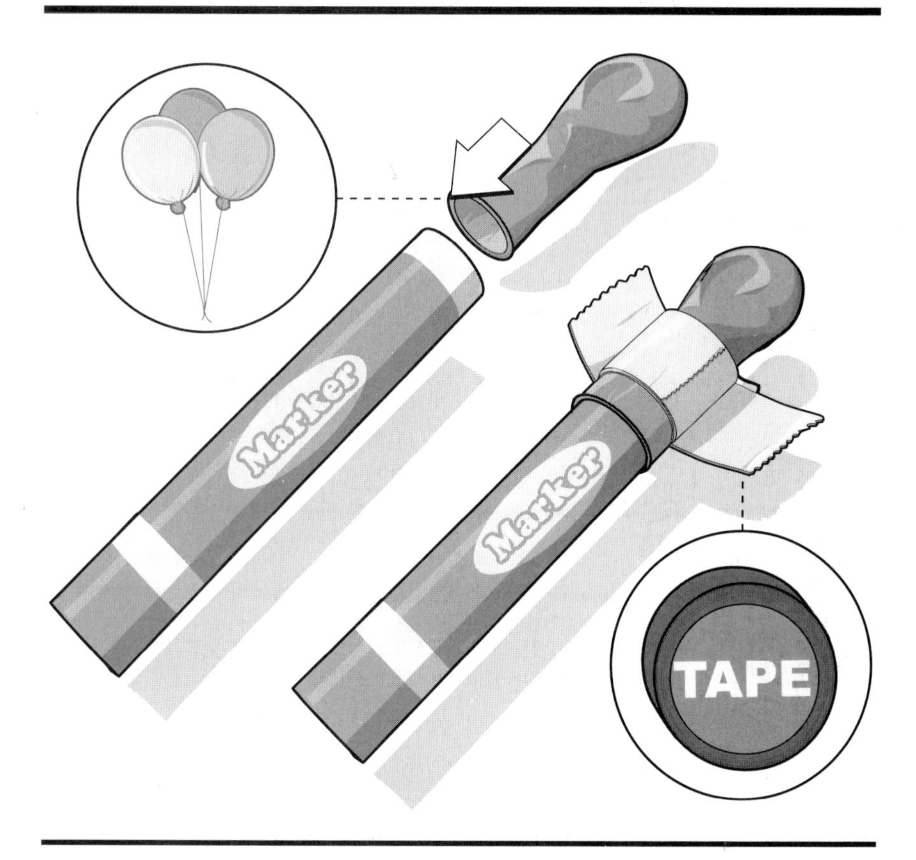

Now it's time to manufacture the power source of your PPK gun barrel. Pull a balloon over the end of the marker housing with a majority of it covering the housing, as shown. Once in place, tape the balloon securely into position. Test the strength of the tape restraint a few times by pulling the balloon, and add more tape if needed. It is important that you test it before you continue construction.

Step 3

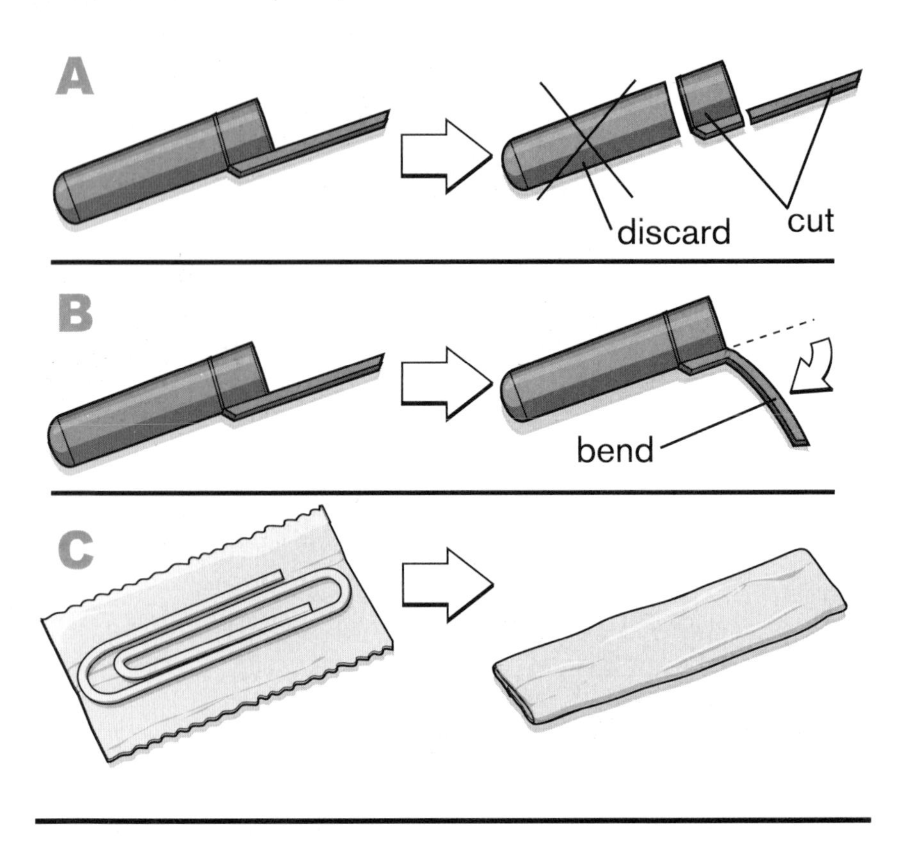

In this step you will prepare three different components that will be used in the next few steps. Round up two plastic pen caps, one large paper clip, and duct tape.

Component A: Using a hobby knife, cut off the pen clip. This will become the barrel sight. Then, from the same pen cap, cut off a ½-inch ring. This will become the hammer detail.

Component B: With the second plastic pen cap, slowly bend down the clip without breaking it. This component will eventually become the trigger.

Component C: Place the large paper clip on a slightly larger piece of duct tape. Securely wrap the clip so that all of it is covered with tape. This is the beginning of the trigger guard assembly.

Step 4

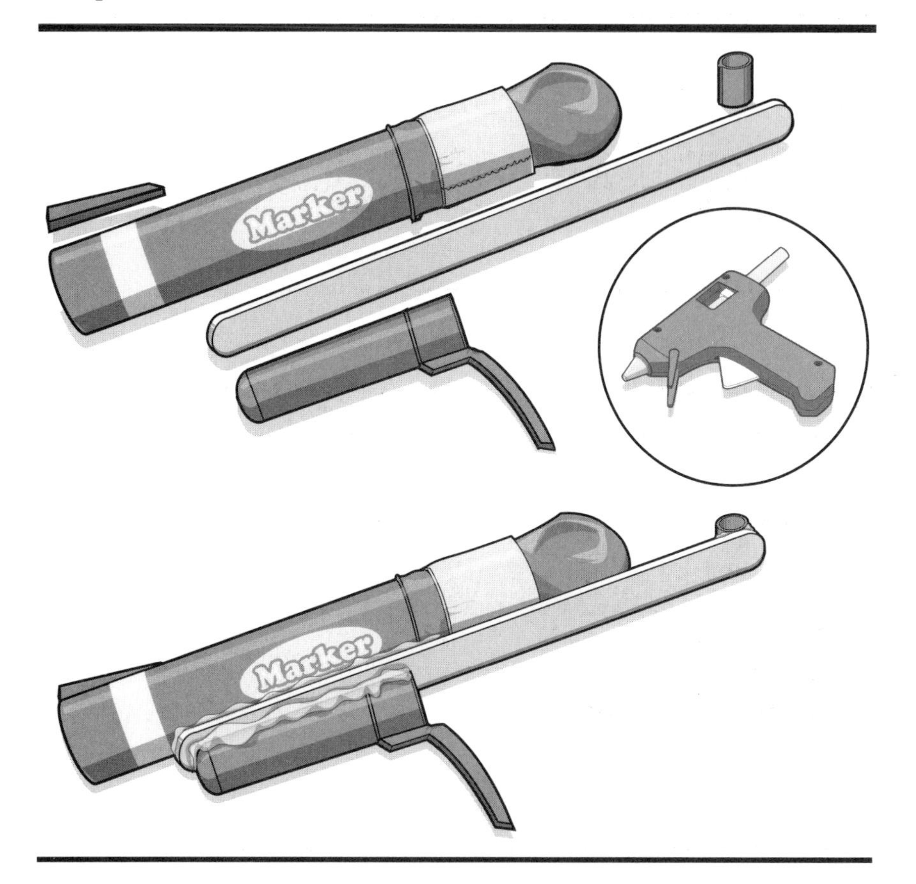

Plug in the hot glue gun. While it heats up, lay out the components according to the illustration. Hot glue the marker barrel onto the craft stick, being careful not to burn the balloon. The craft stick should extend approximately 1¾ inches off the balloon side. Once that has set, glue the bent pen cap to the bottom of the craft stick as shown. After the craft stick has cooled, attach the cut pen cap clip to the top of the marker barrel, and then glue the cap ring to the rear of the craft stick.

Remember to avoid burning the balloon with the hot glue gun. If you do burn the latex, repeat step 3 with a new balloon.

Step 5

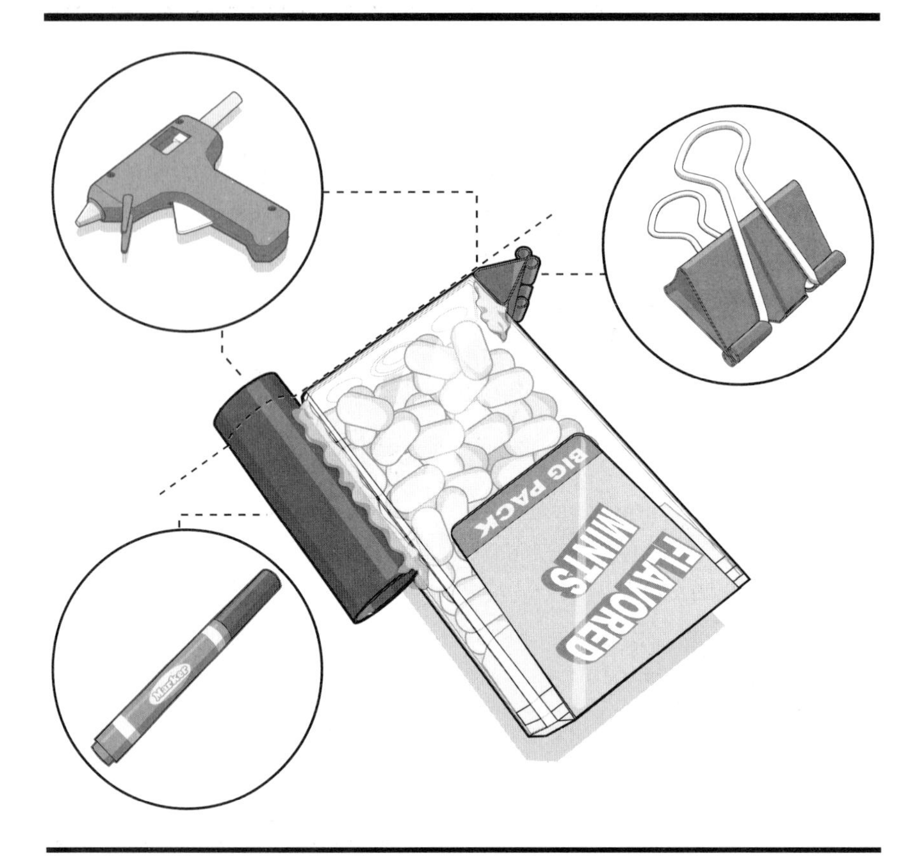

It's time to build the firearm handle and clip assembly. Hot glue the marker cap to the opposite side of the Tic Tac container; the cap should extend from the bottom approximately ¼ inch. Now remove the metal handles of the small binder clip (19 mm) and fix the clip to the opposite side of the pen cap. The clip should be level with the bottom of the plastic container, as shown. The stock assembly has been completed.

Step 6

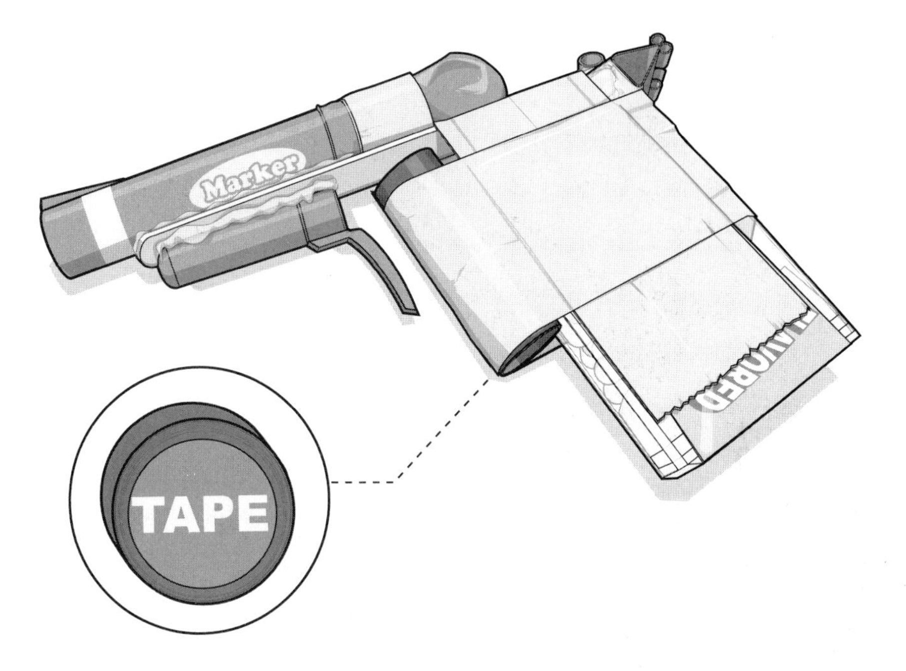

It's time to join both of the assemblies. Place the barrel assembly onto the Tic Tac container so that the craft stick touches the small binder clip. Once in position, use duct tape to secure both items together. Hot glue is suitable, but the duct tape will camouflage the recycled components.

You now have a functioning launcher. The rest of the steps are for aesthetic purposes.

Step 7

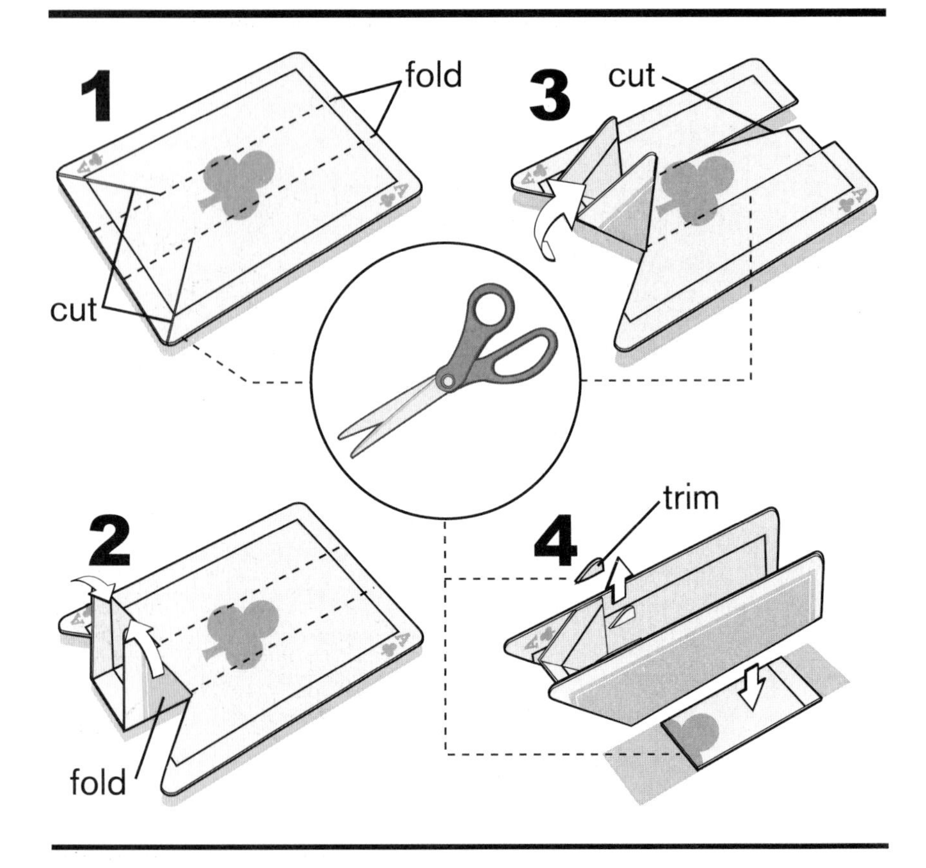

Use two playing cards to transform the PPK Tic Tac's housing. Start with the first playing card.

In the length (longest) direction, with the playing card faceup, fold the card into three sections, with the center section approximately the width of the marker barrel. With scissors, cut two 45-degree incisions starting at the card corner and ending at the two fold lines, as shown.

Fold up both triangles, using the center fold lines as their base.

Toward the center of the card, fold both triangles at a 45-degree angle. On the opposite end of the card, cut two more 1¼-inch straight lines along the center fold lines.

Now fold the sides of the playing card up to align with the cut angle. Once in place, glue the triangles to the side of the card. With scissors, trim the tips of the triangles and cut off the bottom rectangular flap.

Step 8

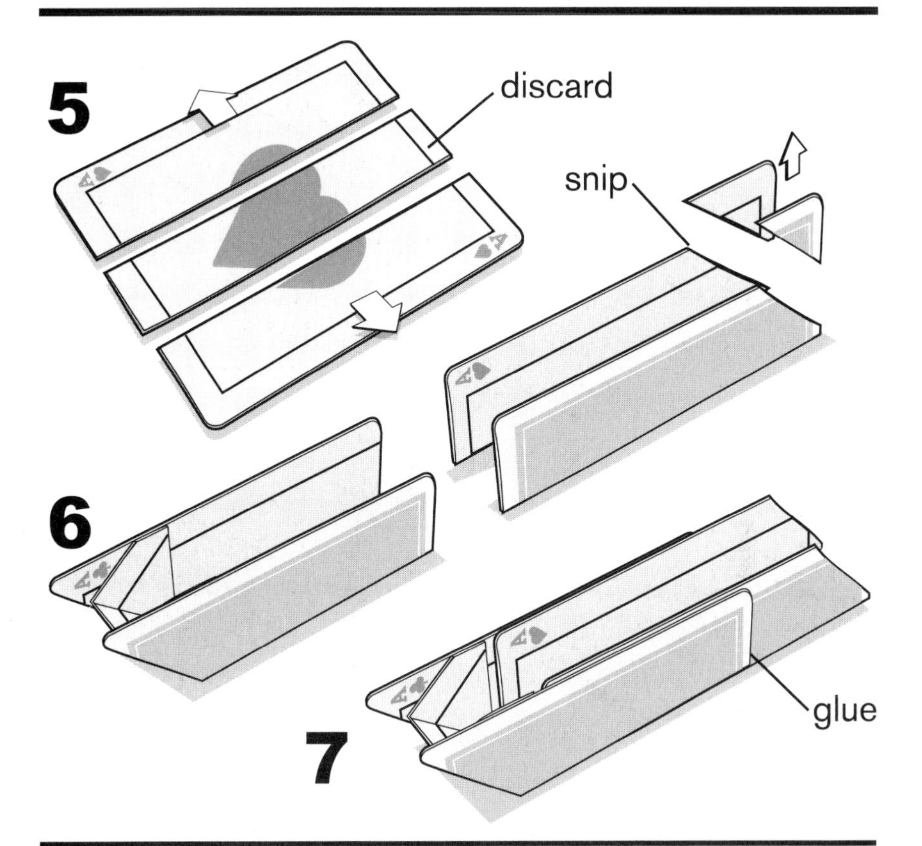

Now modify the second playing card.

Fold the second card similar to the first, making three sections and two crease lines. With scissors, cut along each crease line; you will end up with three sections of card. Discard the center section.

Using the scissors again, snip off one corner of each card section. This will make it easier to access the balloon when shooting and is close to the design of a real Walther PPK, James Bond's favorite gun.

Time for more glue! Attach both sections to the inside of the modified first card. The new sections should protrude 1½ inches, with the total assembly looking like a rhombus.

Step 9

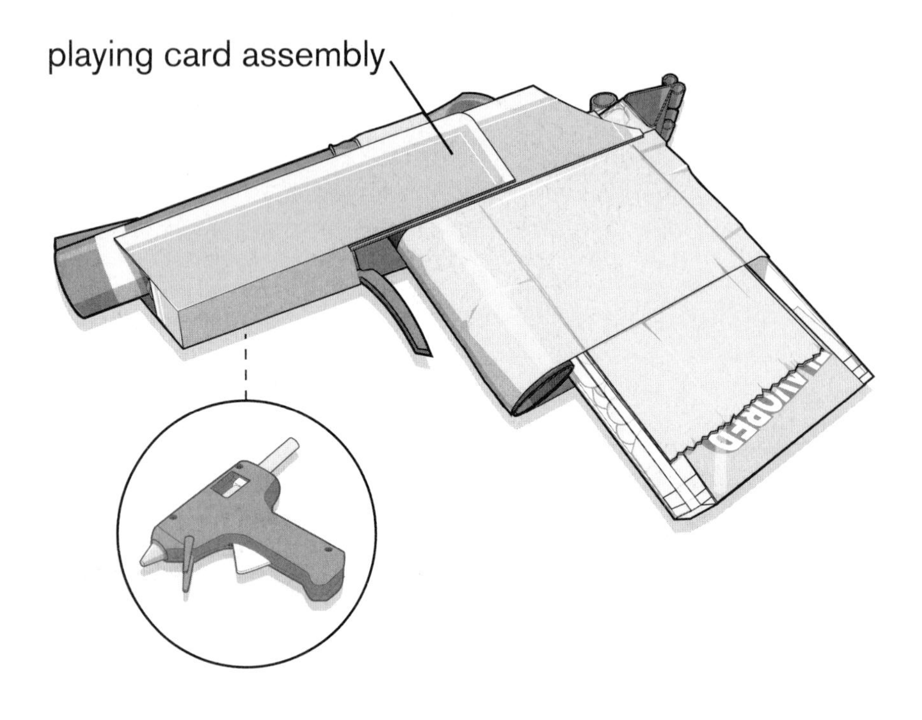

Now slide the playing card assembly underneath the barrel assembly and fasten it into place using hot glue. If the pen cap trigger detail causes a problem with the fitting of the card assembly, just cut off some card material with scissors.

Step 10

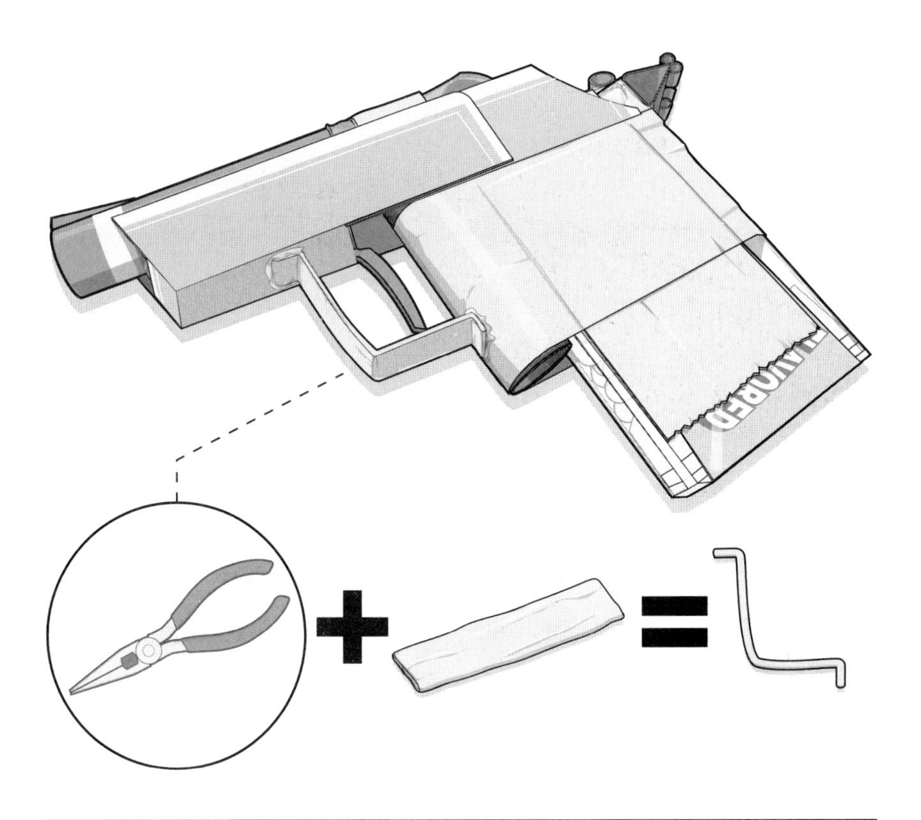

Finally, don't forget your trigger guard. Grab the duct tape–wrapped large paper clip from step 3 and a pair of pliers. Use the pliers to slowly bend the paper clip three times to shape the PPK's trigger guard. Follow the illustration above to determine the bends. Once bent, position it in front of the pen cap trigger and use hot glue to secure the guard in place. Additional tape can also be used.

Now, load a Tic Tac or other small, hard candy into the barrel so that it falls into the balloon. (You can also use a cotton swab, bean, small eraser, or peanut.) Locate it with your fingers. Once you have a grip on it, pull it back, point your shooter, and release. ***Remember: it is important that you pick a safe target to practice your marksmanship.*** Never operate the launcher if the balloon is showing signs of wear; it is possible for ammunition to eject from any tear in the latex.

After you fire, the balloon might get wedged into the marker housing. If it does, just blow it out from the other end of the barrel.

Alternate Construction

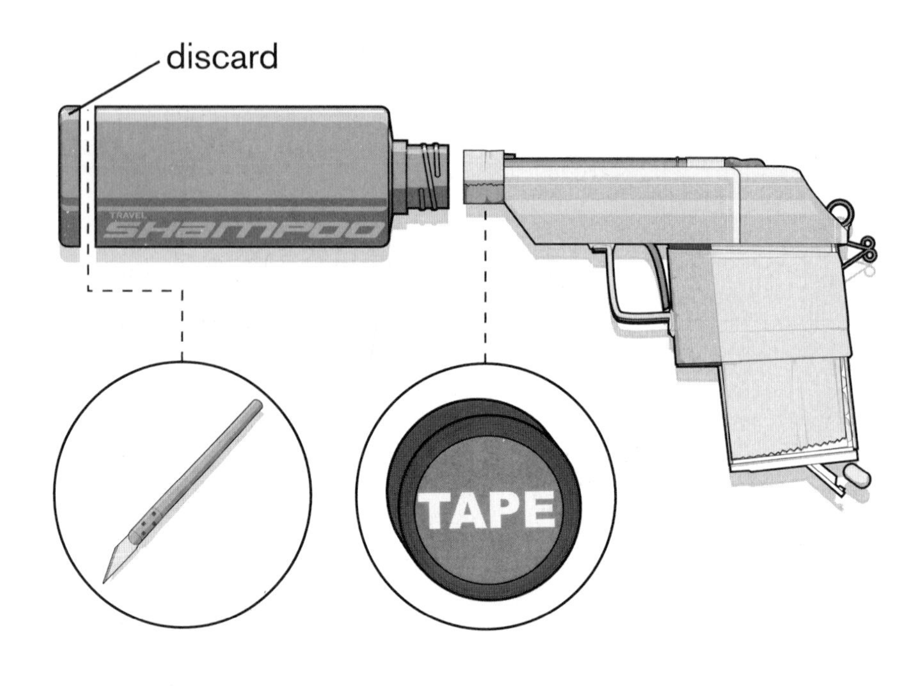

Every Tic Tac PPK needs a highly sophisticated silencer attachment. Raid the bathroom for a travel-sized plastic shampoo bottle. Using a hobby knife, create a small incision at the bottom of the container—then with the blade still in the bottle, slowly rotate the bottle 360 degrees. ***Be careful and take your time—if you rush it, this can be hazardous.*** Use scissors to level or clean up the edge of the cut.

To get the PPK barrel to match the diameter of the shampoo bottle, wrap tape around the barrel tip to increase its diameter. Now slide the silencer onto the barrel for a snug fit, but do not glue it in place. Permanently attaching the silencer will make it difficult to load your launcher.

CANDY GLOCK 33

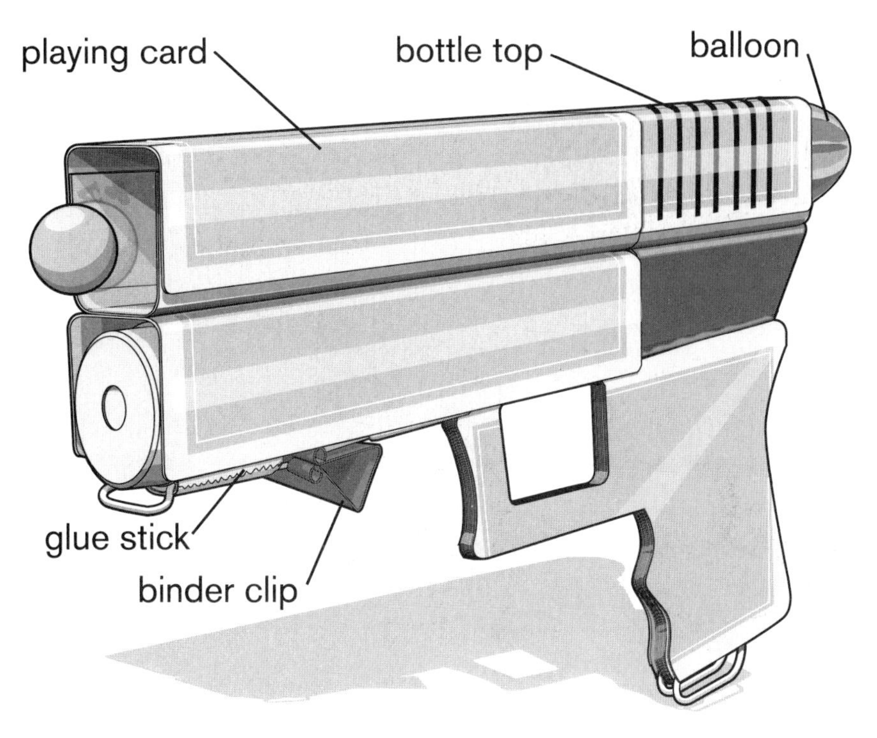

Range: 10–60 feet

Forged out of playing cards, the Candy Glock 33 has a simple and sleek design. It has an innovative flip-down magazine under the barrel, making candy reloading quick and easy. It also features an ergonomic grip that can be customized to the user's hand and an integrated internal metal frame.

Supplies

1 plastic soft drink bottle, approx. 16.9 fl. oz. (500 mL)
1 medium balloon
Duct tape
1 glue stick (0.28 oz. preferred)
1 standard art marker cap
12 playing cards
1 small binder clip (19 mm)
1 large binder clip (51 mm)

Tools

Safety glasses
Pocketknife
Scissors
Marker
Hot glue gun

Ammo

1+ round candies

Step 1

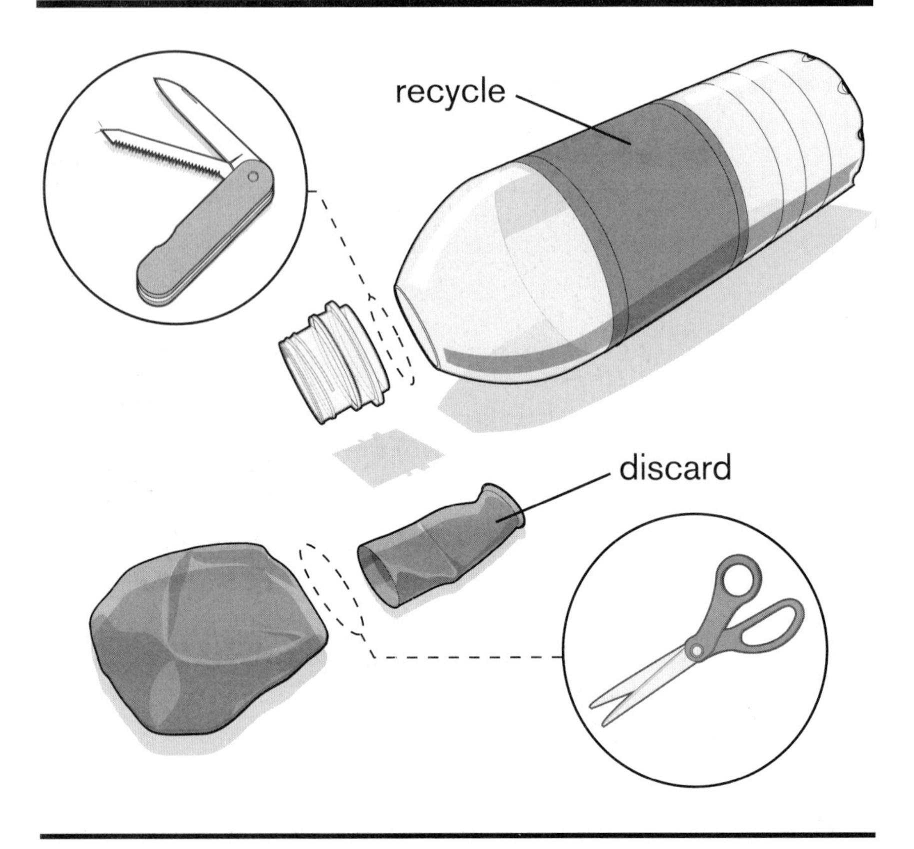

Operation "Candy Glock 33" is a go! The first step is to assemble the bean shooter, the firing mechanism. Using a pocketknife blade, carefully saw or cut off the threaded neck of a soft drink bottle. Once you have removed the neck, use the knife to trim any sharp protrusions that may be left on the cut edge.

Next, use the scissors to cut the balloon in half as shown above. Discard the mouth end.

Step 2

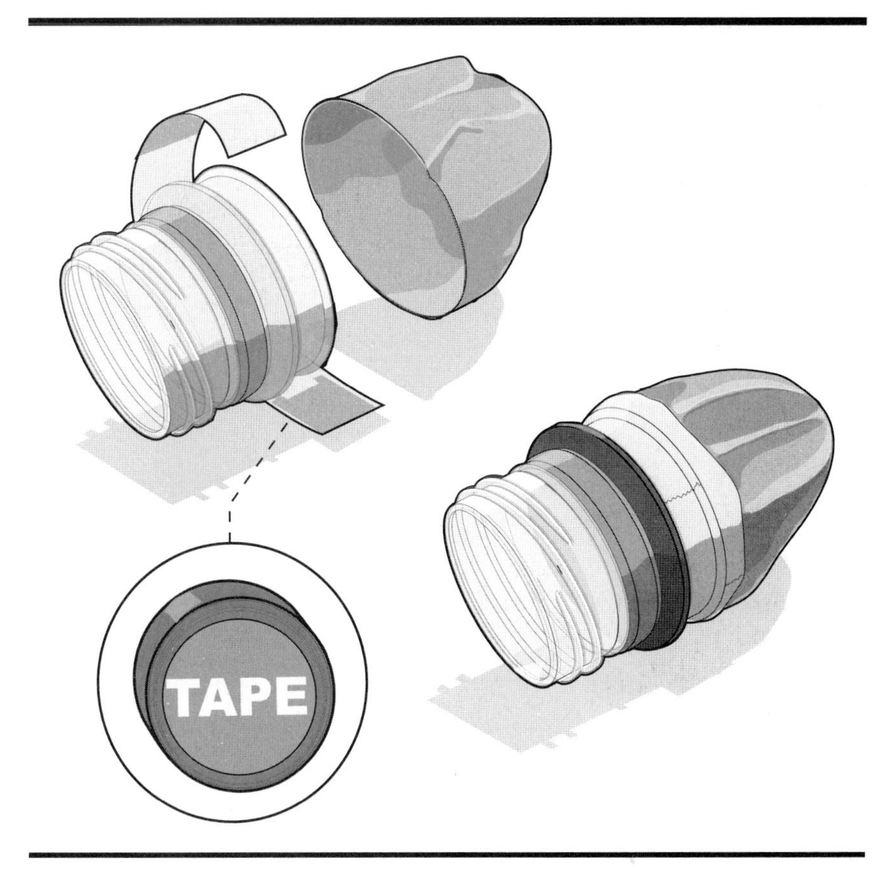

Next, you will tightly tape the half-section of the balloon to the bottle top—but before you do, be certain that you've cleaned up the edges of the bottle in step 1. Sharp edges on the bottle top will eventually lead to holes or rips in the balloon, causing ammunition to fly out unexpectedly.

Your basic bean shooter is complete. Feel free to test it out. Drop a bean or piece of candy ammo into the shooter, pull back on the balloon, aim it away from any person or pet, and let it fly.

Step 3

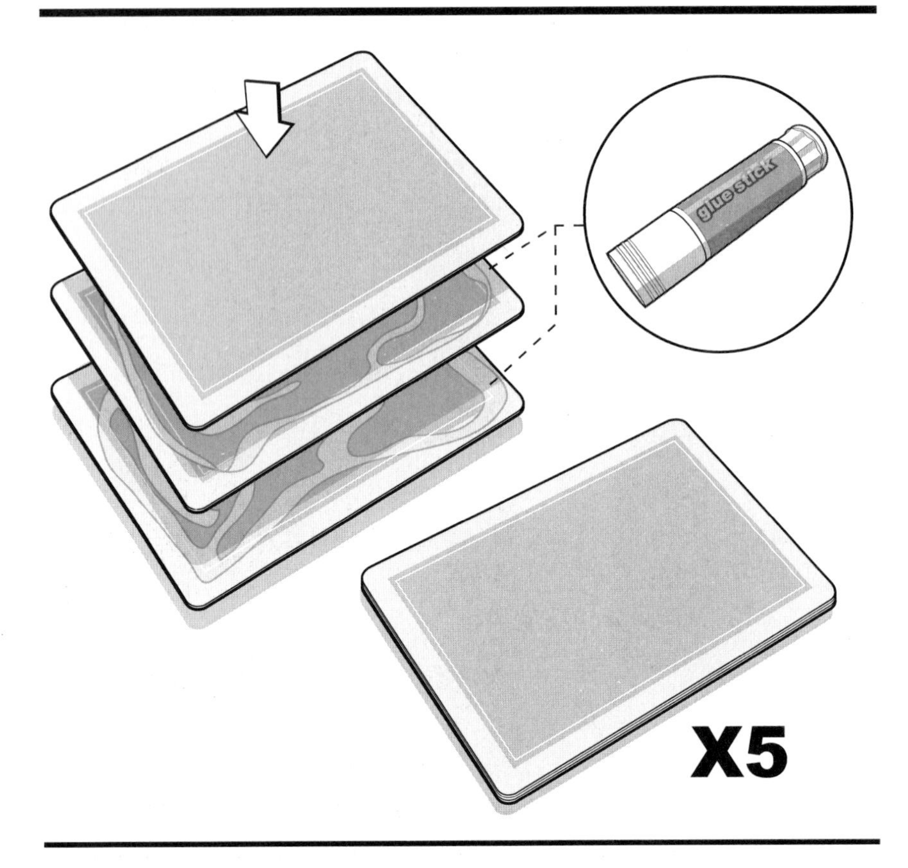

Next, you'll start the assembly of the Candy Glock 33's stock. Using a glue stick, sandwich five playing cards together in a perfect stack, then let the cards set until they are completely dry.

Step 4

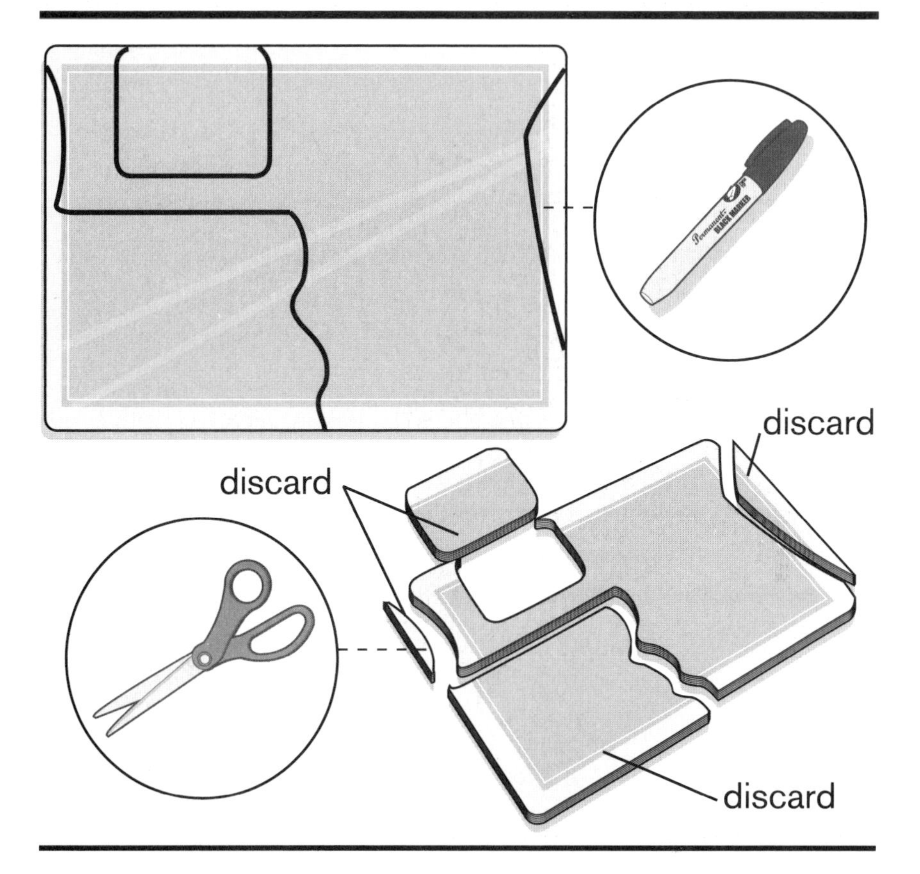

Using a marker, draw the suggested grip pattern shown above onto the stack of cards. The suggested pattern includes finger grips and a trigger guard opening that should be drawn large enough for your finger to be comfortable.

Use scissors to cut the pattern out. Discard the scraps.

Step 5

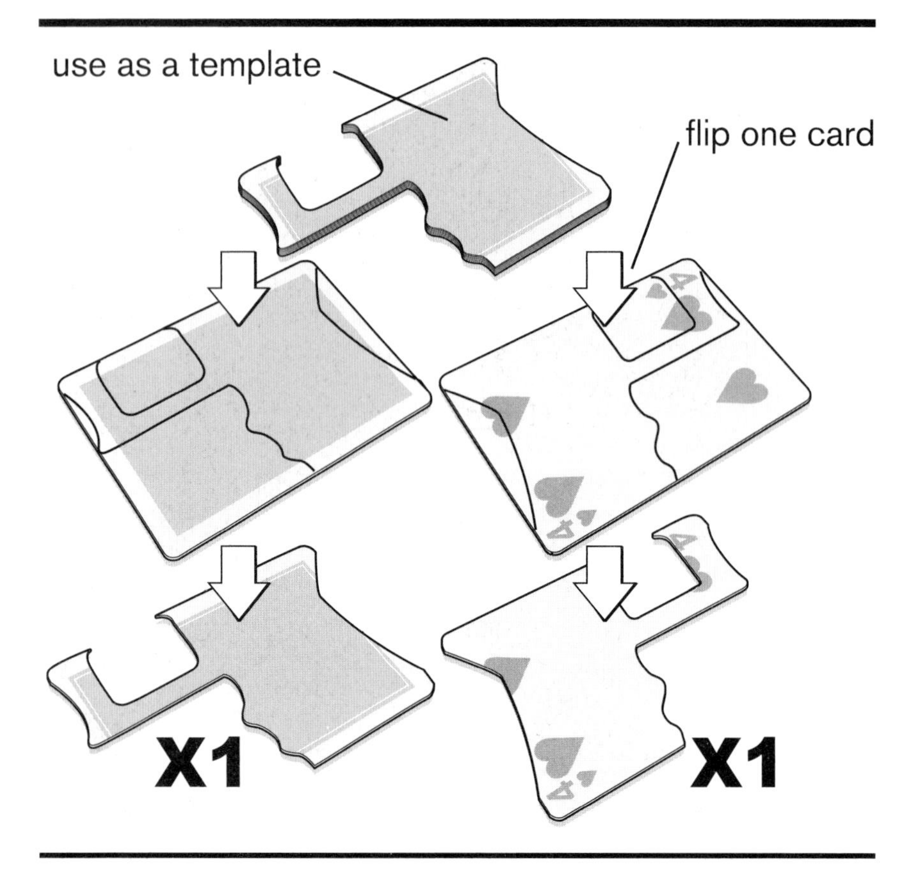

In this step, use the five-card assembly as a template for two more individual cards.

Align the template on top of one card and then use a marker to trace the outline. With scissors, remove the pattern and discard the scraps.

Draw the pattern on the reverse side of the second card and again remove the pattern with scissors. Drawing the pattern on the reverse side is for visual reasons only; it will insure that only the card back will be visible on the final gun, and not the faces of the card.

Step 6

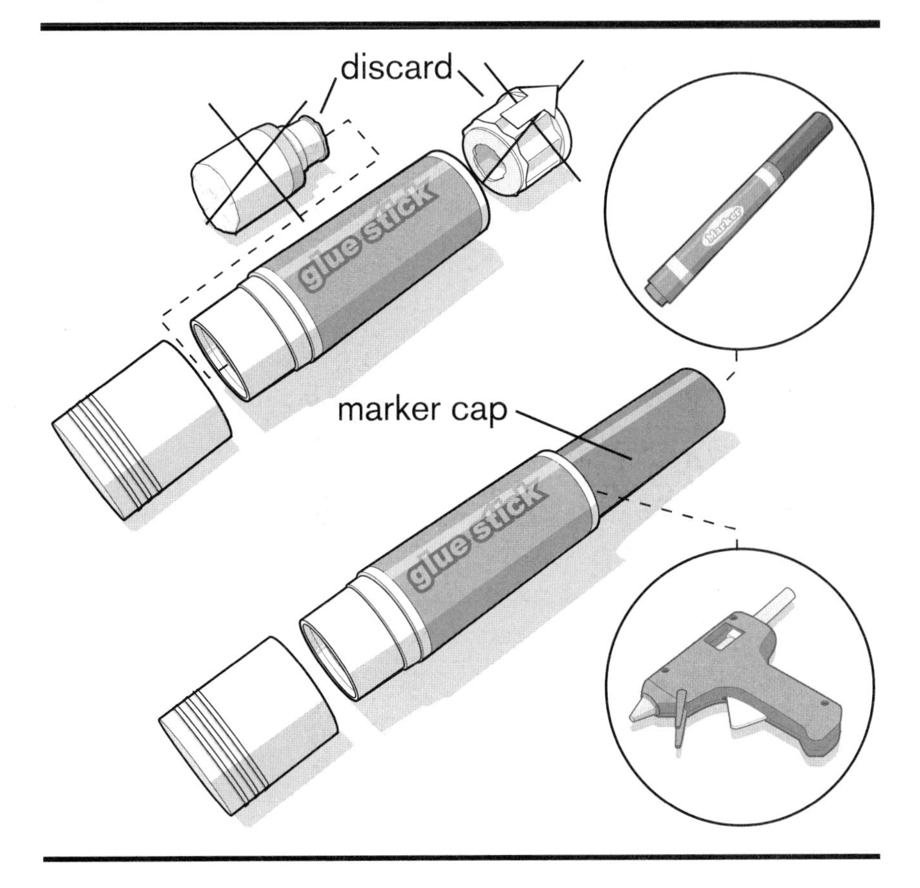

The next step is the assembly of the flip-down magazine chamber, which is optional but convenient. Although a 0.28 oz. (8 g) glue stick is preferred, if this is not available, a lip balm container or larger glue stick (0.77 oz.) can be substituted. Remove the glue from inside the glue stick container by snapping off the rotating bottom end and pulling out the innards.

Now, hot glue the cap from a standard art marker to the bottom end of the glue stick container. Later, in step 9, this cap will fit snugly inside a large binder clip.

Step 7

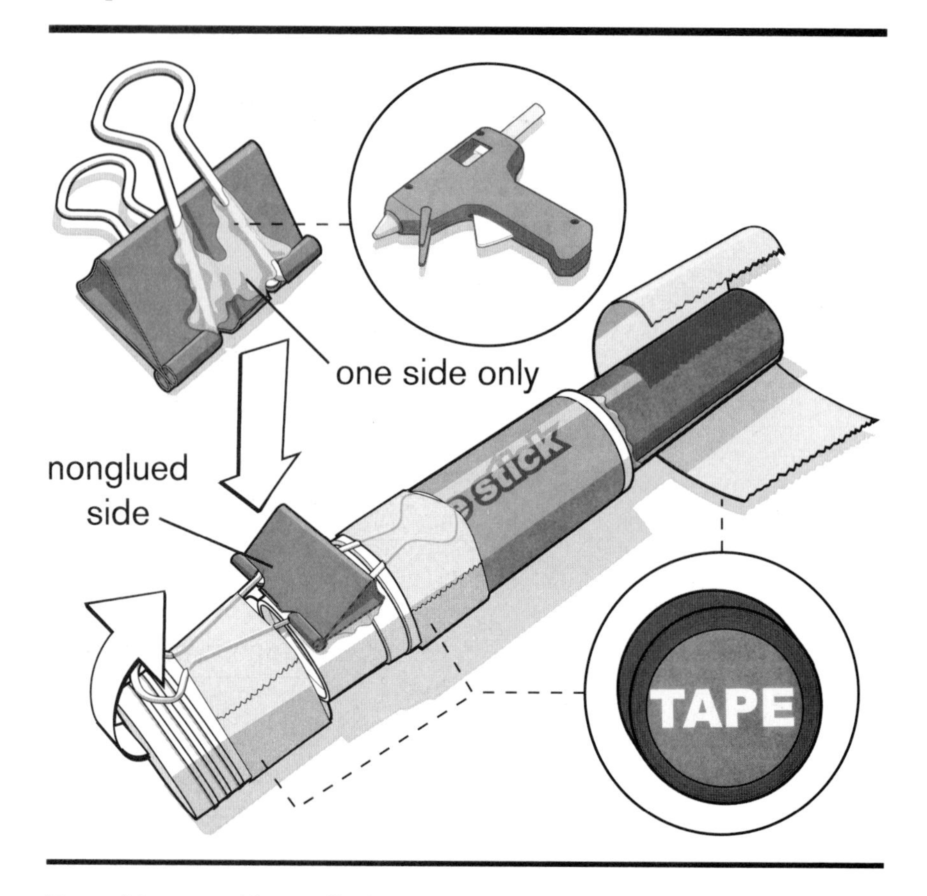

To avoid ammunition spills from the magazine, add a hinge system that is fabricated from a small binder clip. Before attaching the binder clip to the container, you will need to hot glue the metal handle on one side only.

Now place the modified clip at the end of the glue stick, with the glued handle facedown, and tape the binder clip into position as shown. Then place the cap in front of the container and tape it to the binder clip metal handle. Once assembled, the cap should "click" and lock up and down.

Finally, wrap 11 inches of duct tape around the glued marker cap to increase its diameter.

Step 8

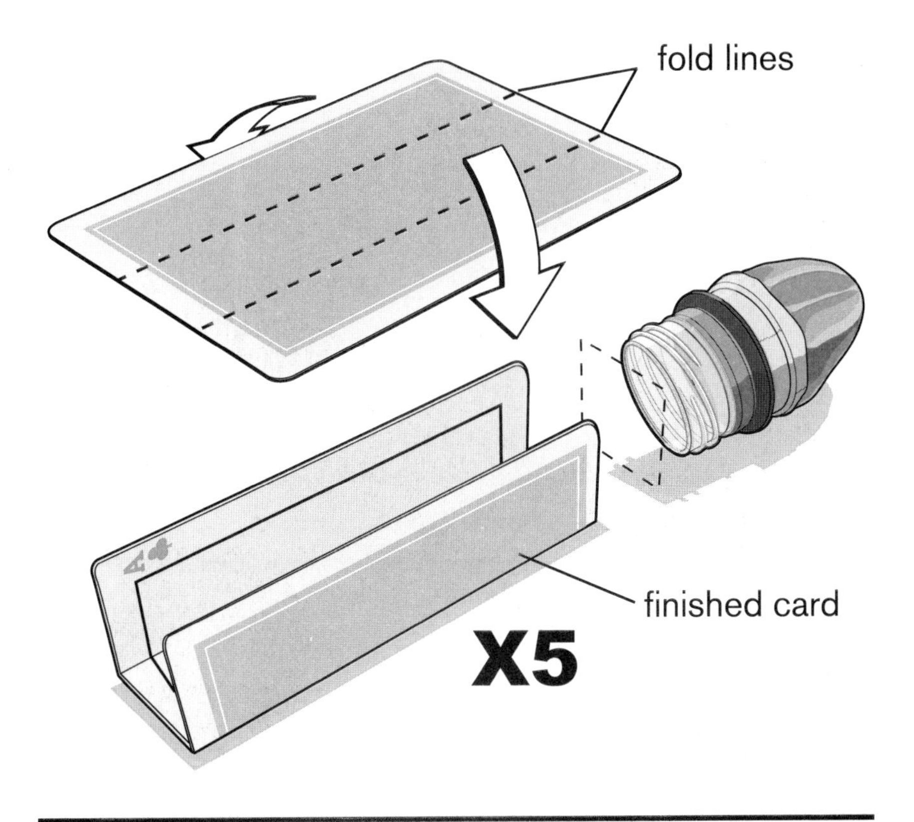

The barrel on the Candy Glock 33 is constructed from playing cards. You will need to fold each card twice; the center width will be the diameter of the bean shooter. Use the bean shooter as a guide and fold the two sides of the playing card at a 90-degree angle.

Put crisp crease line in the folds, and then repeat this step with four additional cards (for a total of five folded cards).

Step 9

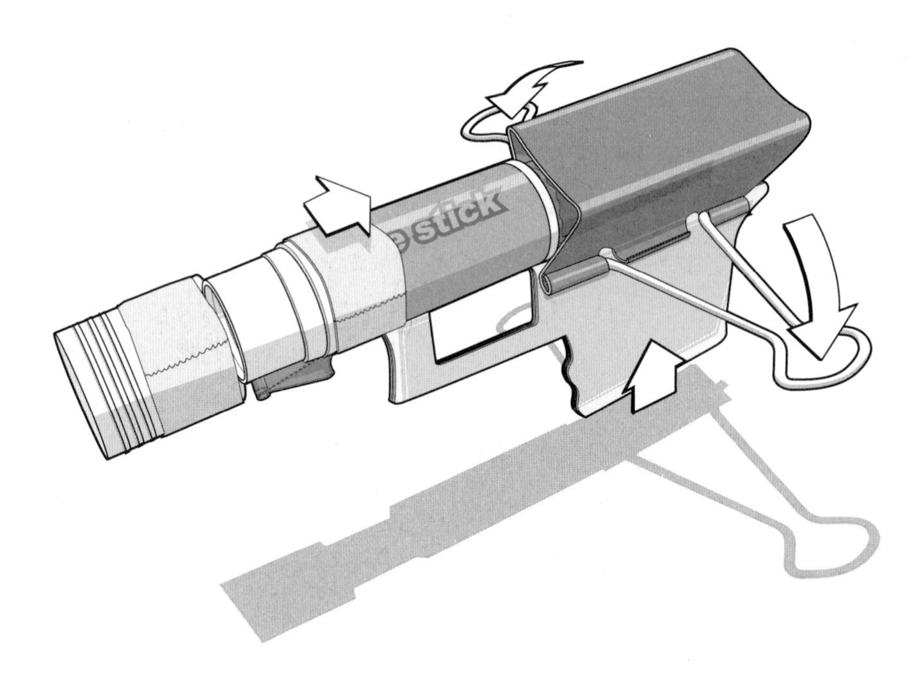

It's time for the Candy Glock 33 to take shape. Start by sliding the marker cap end of the magazine clip assembly into the triangular void of the large binder clip. The 11 inches of added tape around the marker cap should make for a snug fit, but adjust if needed. The binder clip hinge should be under the magazine, as shown.

Next, place the five-card custom grip approximately ½ inch into the large binder clip and lock it into place. Flip the metal handles down when finished.

Step 10

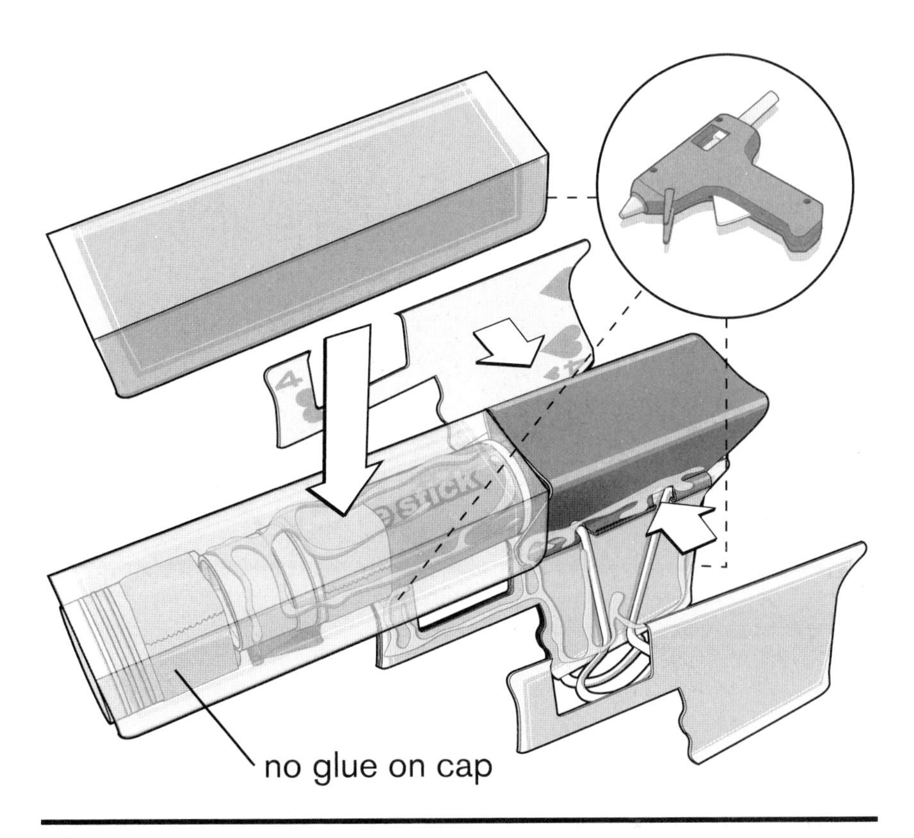

Glue one of the folded playing cards onto the glue stick container, but be careful not to glue the card onto the hinged cap.

Next, coat one side of the playing card handle with hot glue and then place one of the single grip cards over the metal handle. Repeat this step for the opposite side, with the last grip pattern card.

Once both cards are fixed to the grip, add a little hot glue between the trigger guard and glue stick magazine to bring it all together.

Step 11

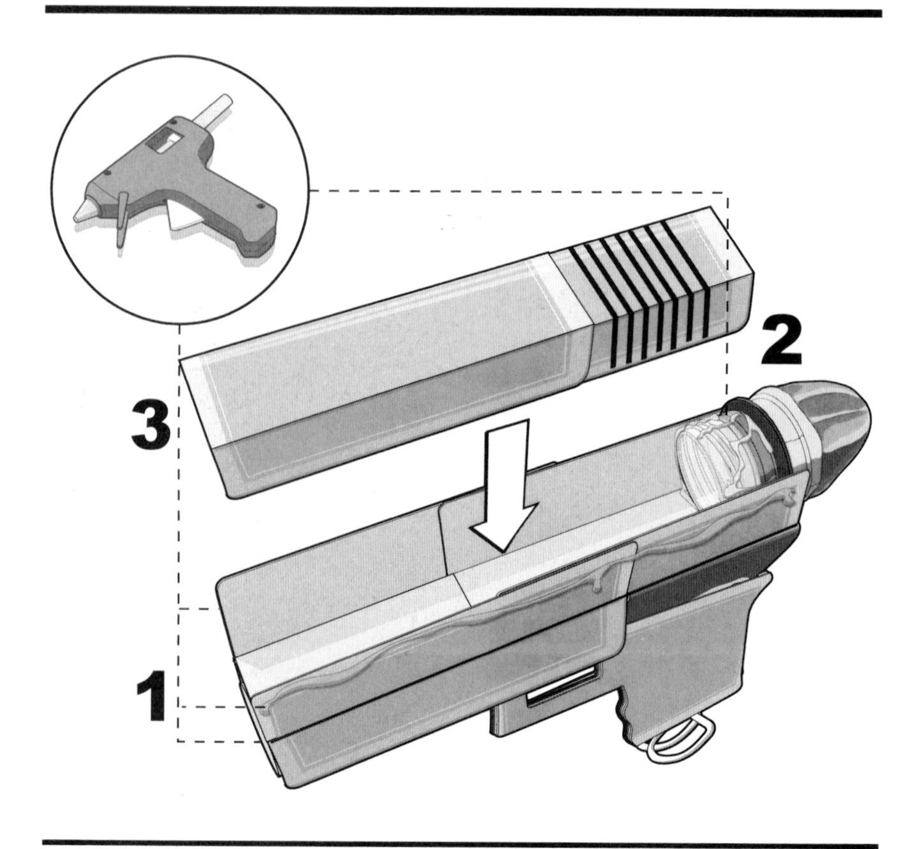

The barrel assembly is the last step. First, hot glue two of the folded cards faceup to create a channel. These two cards should be flush with the existing fixed card and large binder clip. You will need to overlap the cards to strengthen the barrel support.

Second, once the channel is in place, heavily glue the bean shooter assembly into the rear of the channel as shown. The balloon firing mechanism should be protruding from the back, making it easily accessible to grip.

Third, use the final two cards to cap the barrel assembly and complete the chamber. Again, add glue to both the existing cards and the bean shooter assembly for support.

Now load a round candy—or an eraser, mini marshmallow, or peanut—into the barrel and locate it with your fingers. Once you have a grip on it, pull it back and release it. It is important that you pick a safe target. This launcher shoots with incredible force!

COTTON SWAB .38 SPECIAL

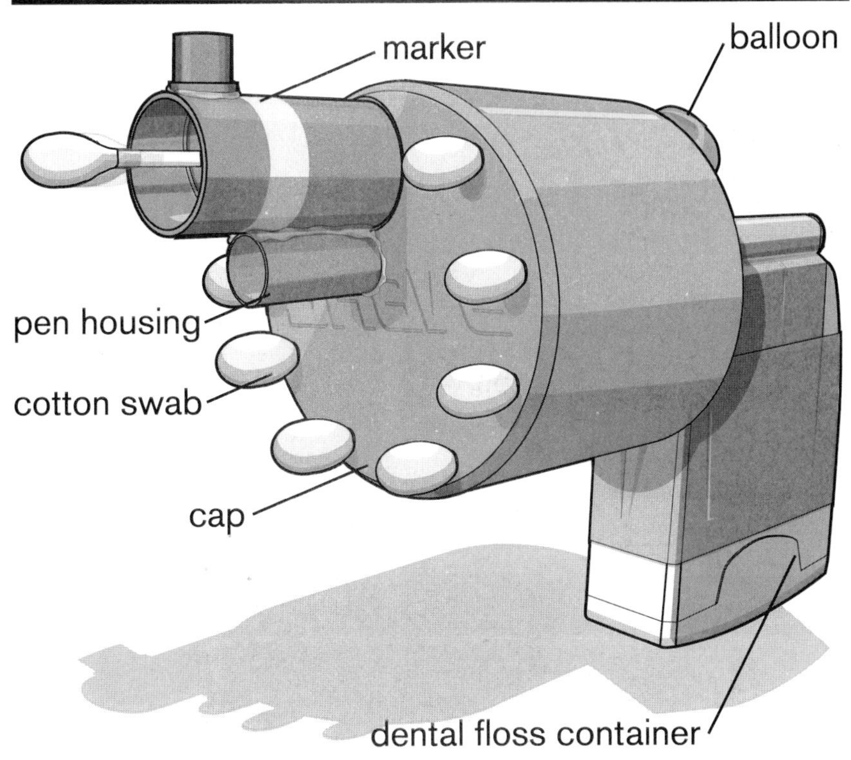

Range: 10–60 feet

Recycle–Reuse–Reload! The Cotton Swab .38 Special is manufactured with lightweight recycled plastic. The friendly design is smooth to operate and extremely accurate. With a small barrel designed to reduce ammunition jams, it's the perfect handheld MiniWeapon starter.

Supplies

1 standard art marker
1 small balloon
Duct tape
1 plastic ballpoint pen
1 large, empty dental floss container
1 plastic shaving cream can cap (approx. 2 inches in diameter by 1¾ inches high)

Tools

Safety glasses
Pliers
Hobby knife
Scissors
Hot glue gun

Ammo

9+ cotton swabs

Step 1

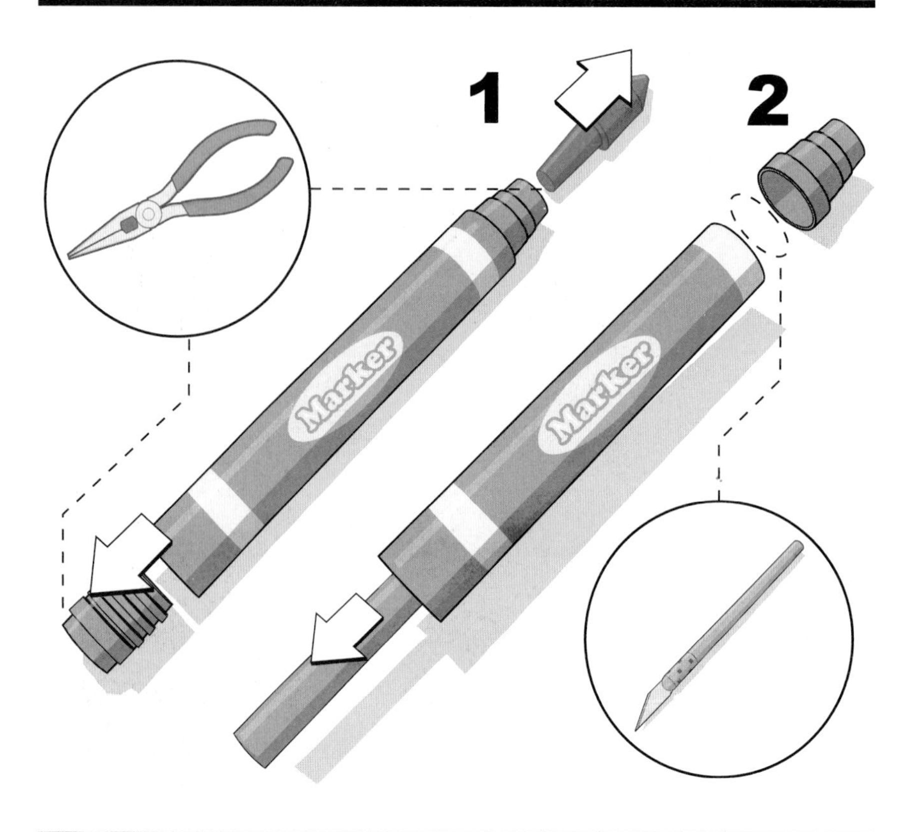

The barrel of this launcher is made out of one standard art marker. A recycled plastic marker is preferred because of the material's softness, but any marker will work. Using pliers, remove the marker's end cap by slightly rotating it from the housing. Again, use pliers to remove the marker nib by pulling it from the opposite end of the marker. Discard the nib to avoid a mess.

Using a hobby knife, pierce the housing below the plastic point on the marker housing and slowly rotate the marker, not the knife. This will make cutting it safer. Once this end has been completely removed, the internal ink cartridge should slide out. Discard this cartridge. (If the housing contains ink residue, rinse it out and dry it.)

The now-empty tube should be unobstructed by any plastic fragments that might have broken off during cutting. Save the marker cap for a later step.

Step 2

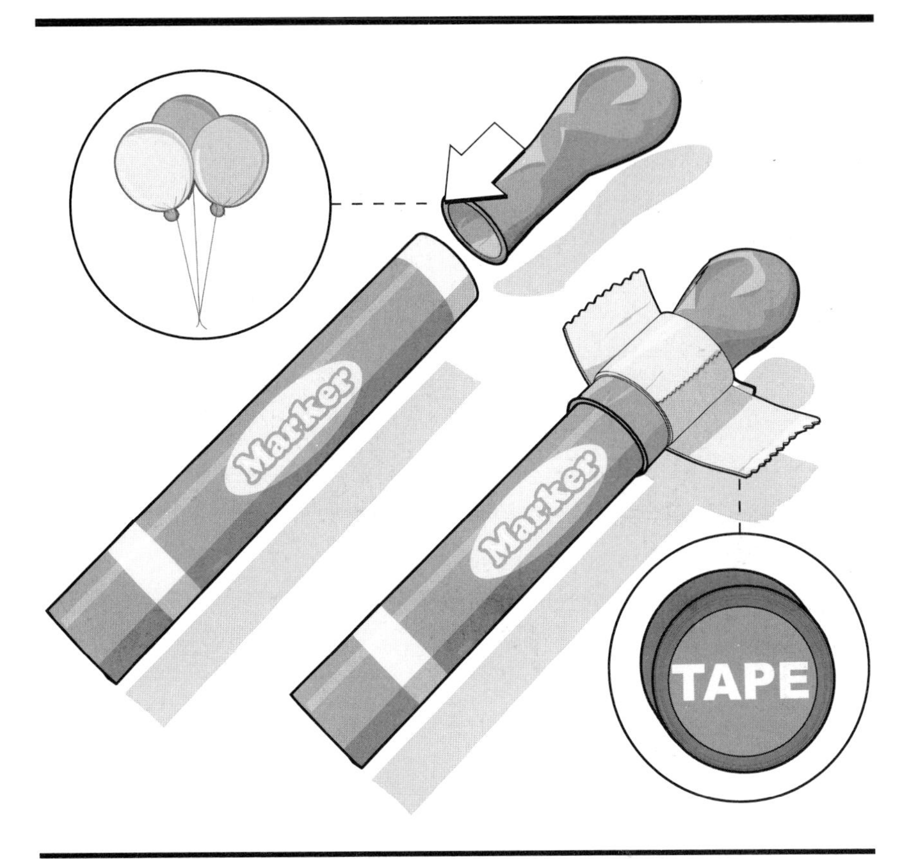

Elastic firepower! Pull a balloon over the end of the marker tube housing; a majority of it should cover the housing. Once in place, tape the balloon securely into position. Test your tape job a few times and add more tape if needed. It is important that you do this before you begin construction on the rest of the MiniWeapon.

Step 3

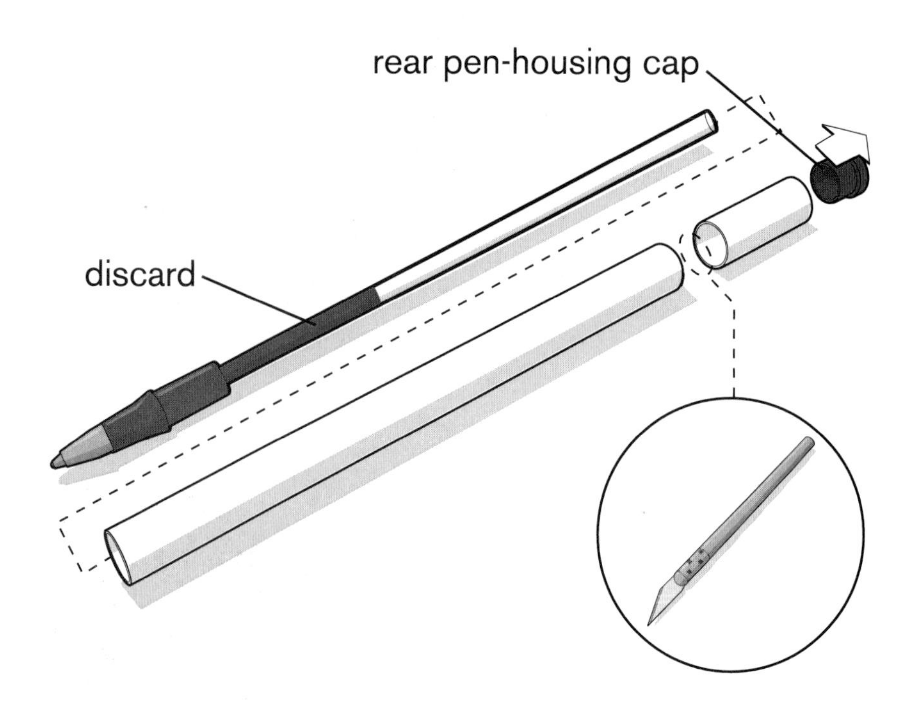

Next, disassemble a plastic ballpoint pen into its various parts. Depending on how the pen has been manufactured, you may need a tool to assist you with dislodging the rear pen-housing cap. A hobby knife (for cutting it off) or small pliers (for pulling it out) would both work just fine. Once the pen is disassembled, use a hobby knife to remove 3/4 inch of material from the pen housing.

Step 4

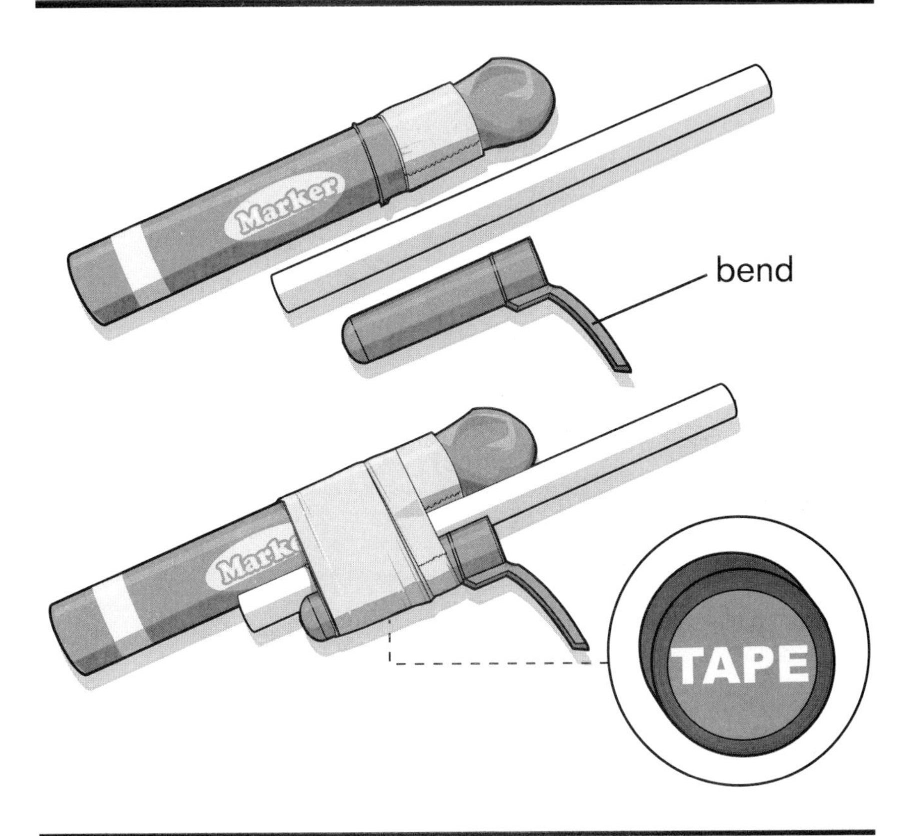

Slowly bend back the plastic clip detail on the pen cap. Then tape the pen housing 1¾ inches onto the marker barrel, protruding from the balloon end as shown. After that is secure, tape the bent pen cap onto the pen housing approximately ½ inch from the end of the pen housing.

These measurements are only guidelines; products and spacing may vary.

Step 5

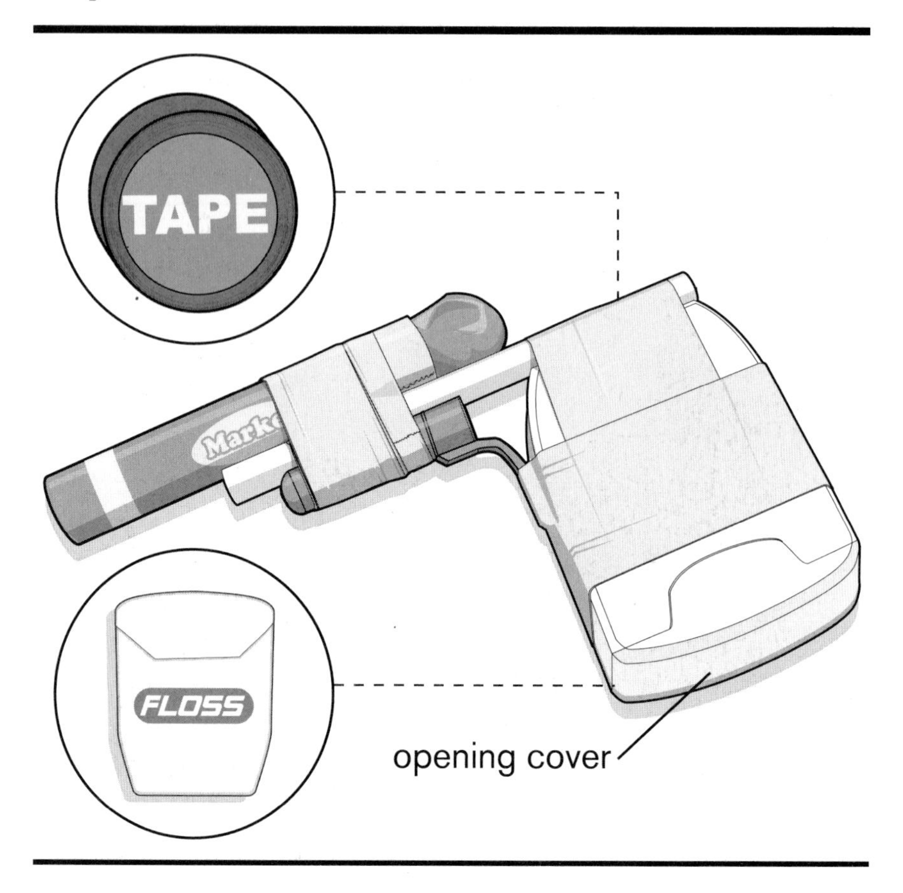

Fix the completed barrel assembly to the bottom end of the empty dental floss container. Apply tape to both the bent pen cap and the empty pen housing as shown, but do not cover the dental floss container's opening.

This container has been transformed into the revolver's handle and also has a nifty little compartment to hold extra ammo.

Step 6

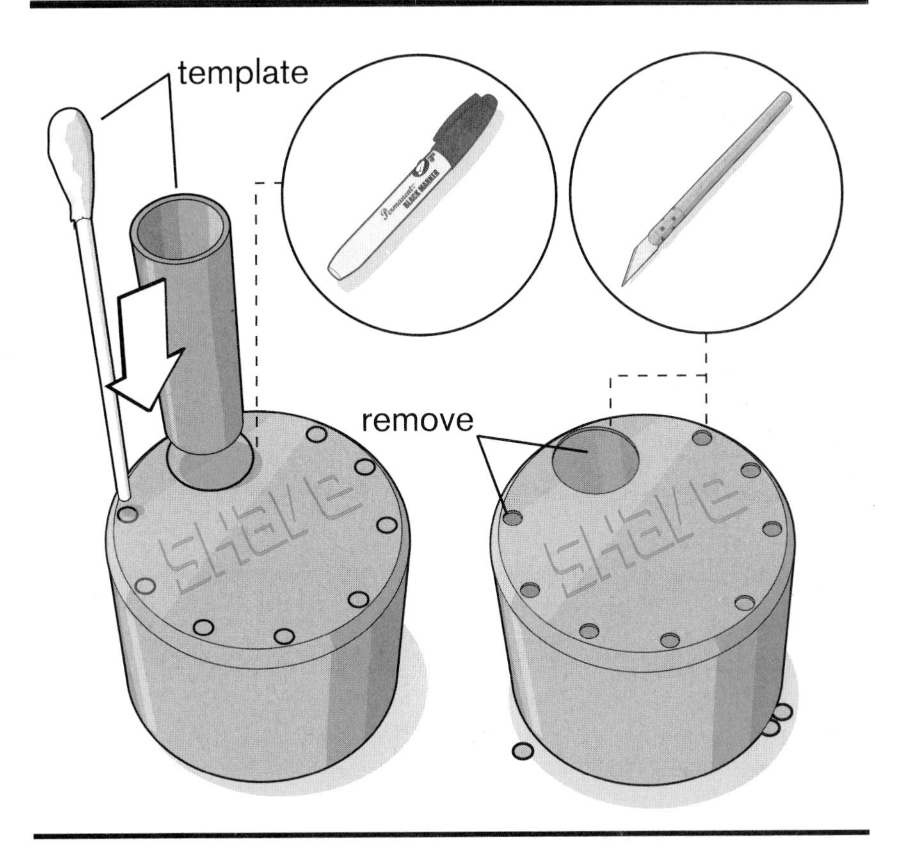

The next step is to create the revolver's cylinder. Use the cap from a small shaving cream container, very similar in size to an aerosol can cap. Different size caps will definitely change the look of this launcher, so feel free to experiment. The barrel will pass through this cap, so you will need to cut a hole through it. To do this, use the marker cap as a template. Trace the marker cap first, near the edge as shown. After that, mark an additional eight smaller sized holes to house the cotton-swab bullets. Remove one end of the cotton swab to determine the desired size.

Now, *carefully* use the hobby knife to slice out the areas you just marked, then test-fit both the barrel and a cotton swab in each hole. Perfect circles will be impossible—just remove enough material to allow the items to fit.

Step 7

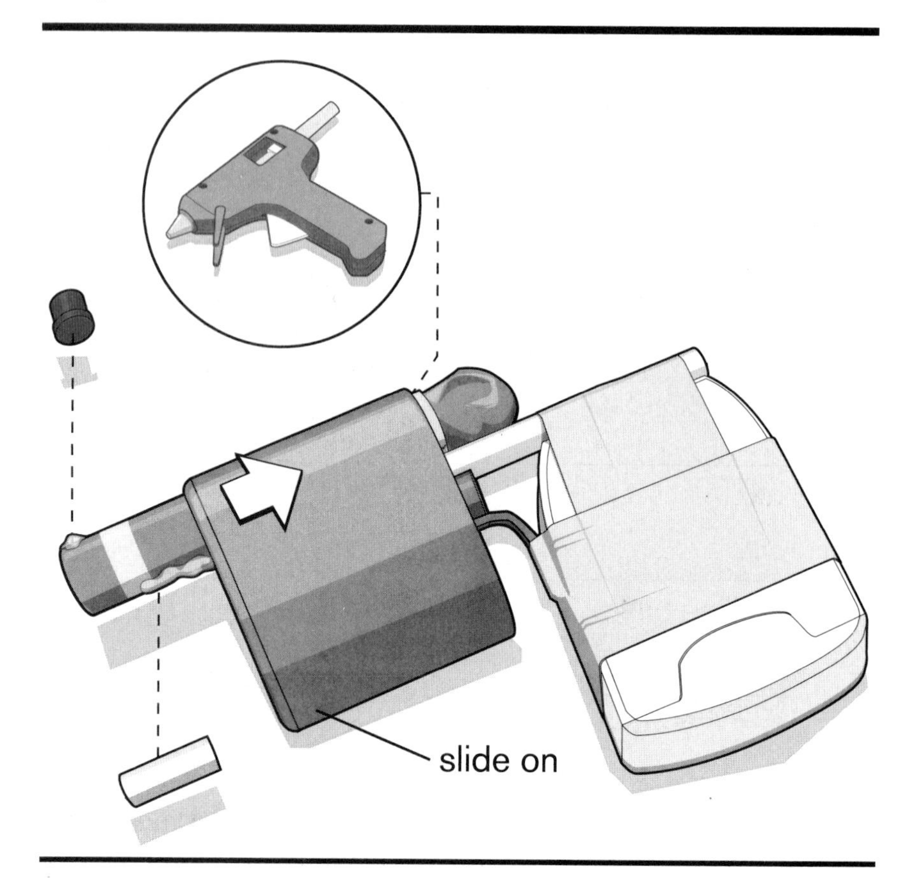

Slide the revolver cylinder over the marker barrel, then hot glue the cap into place. For some finishing touches, glue the rear pen housing to the tip of the barrel for a built-in sight. Then add the short section of pen housing under the barrel to mimic the extractor rod cylinder.

Now cut off the ends of eight more cotton swabs and slide them into the small holes you cut into the revolver cylinder. Any extras, or different types of ammo, can be placed in the dental floss handle.

Lock, stock, and one-marker barrel! To fire, load one of the modified cotton swabs into the barrel, allowing it to drop down to the balloon, with the cotton tip pointing out. Locate the swab's stick inside the balloon with your fingers. Once you have a grip on the stick and a target, pull it back and release.

Although cotton swabs have a cotton tip, they should always be pointed at a safe target.

VEST POCKET MINI

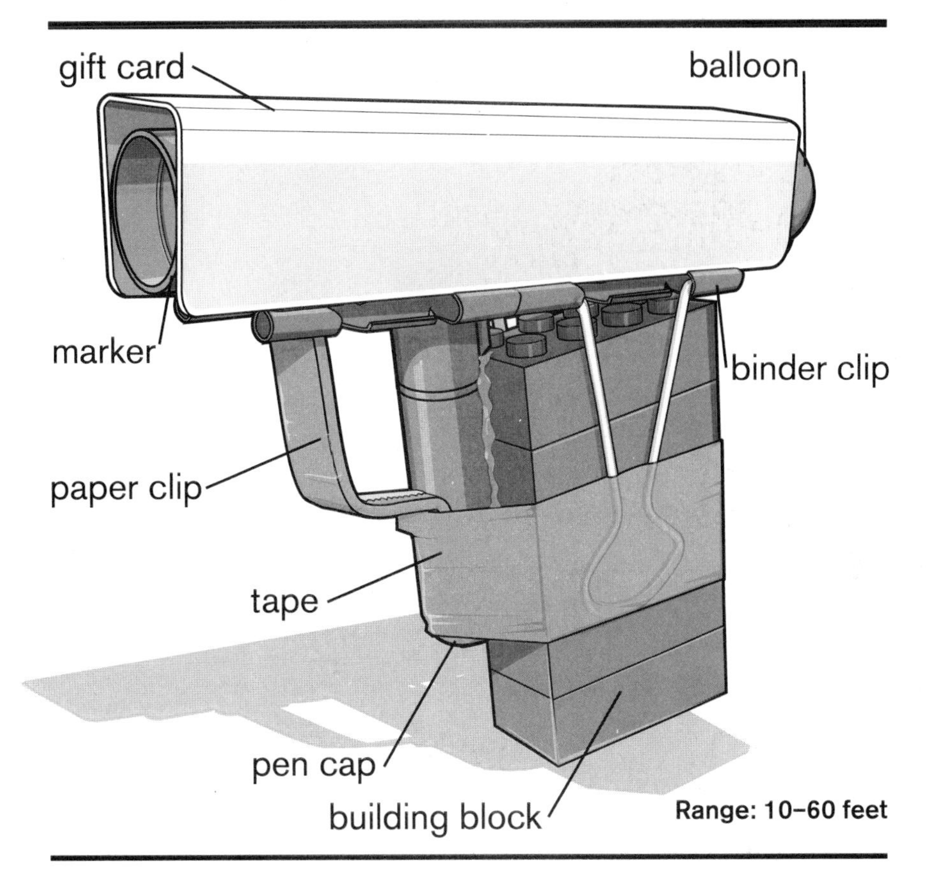

Compact like a butterfly but stings like a bee, the Vest Pocket Mini is a popular secondary sidearm. When it's unleashed, minions and henchmen will be blown away with a single shot. It also features a customizable building-block handle, making it adjustable and easy to carry.

Supplies

1 standard art marker
1 small balloon
Duct tape
1 large paper clip
2 medium binder clips (32 mm)
5 2-peg-by-4-peg building blocks
1 plastic pen cap
1 expired plastic gift card

Tools

Safety glasses
Pliers
Hobby knife
Scissors
Hot glue gun

Ammo

1+ small, hard candies

Step 1

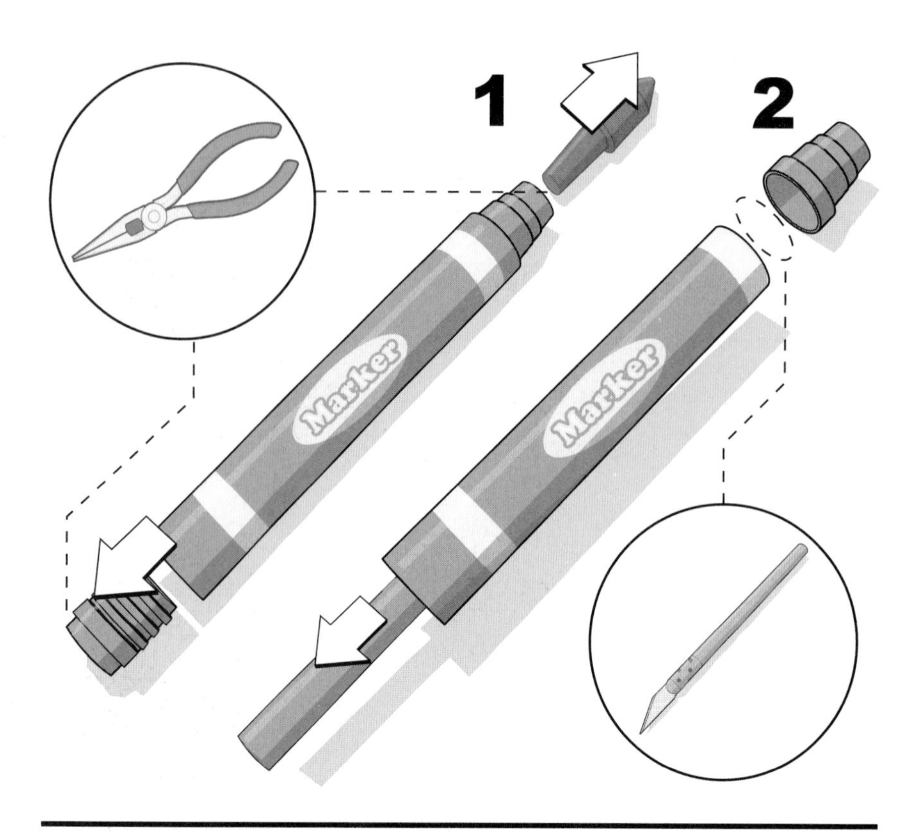

Similar to other sidearms in this chapter, the Vest Pocket Mini is made from one standard art marker. If possible, a recycled plastic marker is preferred because of the material's softness, but any marker will work. Using pliers, remove the end cap by slightly rotating it from the housing. Again, use the pliers to remove the marker nib by sliding it out the tip of the marker. Discard this nib to avoid a mess.

Using a hobby knife, pierce the housing below the plastic point on the marker housing and slowly rotate the marker, not the blade. This will make cutting it safer. Once this end has been completely removed, the internal ink cartridge should slide out. Discard this cartridge. (If the housing contains ink residue, rinse it out and dry it.)

The marker housing pathway should be unobstructed by any plastic fragments that might have broken off during cutting. Discard the marker cap.

Step 2

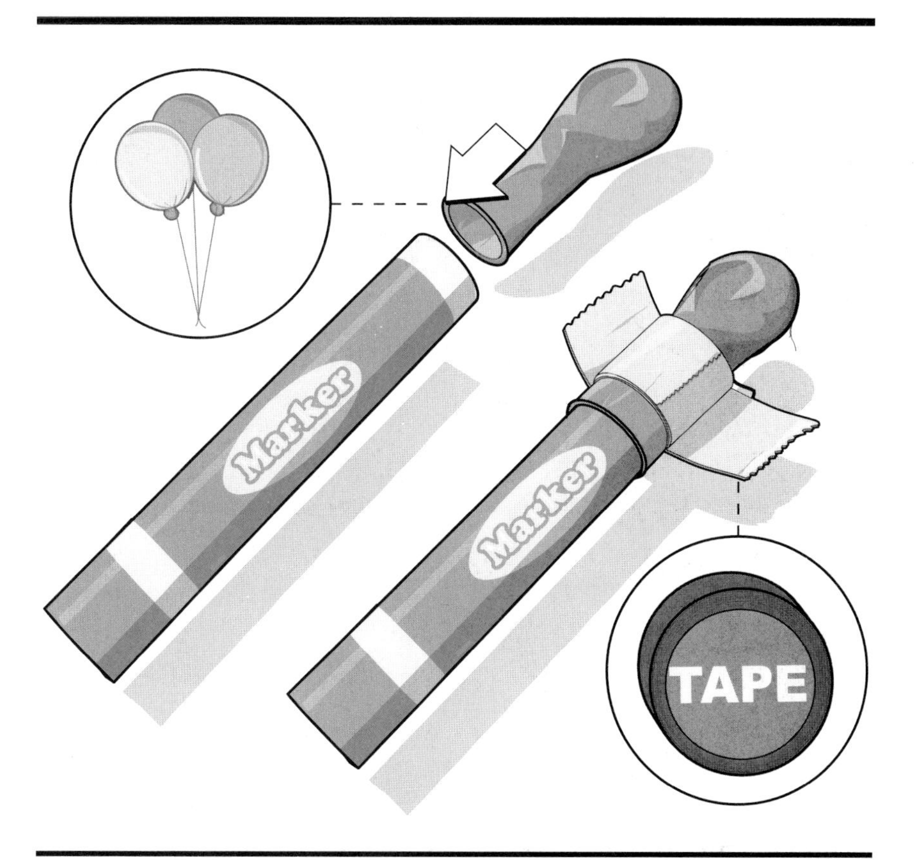

The Vest Pocket Mini sports an unbelievable single-shot firing mechanism, made from a standard balloon. Put a water-balloon sized balloon over the end of the marker housing; a majority of the balloon should cover the housing. Once in place with roughly ½ to ¾ inch overhanging, tape the balloon securely into position. Test your tape restraint a few times, and add more tape if needed. It is important that you do this before you build the rest of the sidearm.

If the balloon is too large, use scissors to remove some of the balloon material near the mouthpiece, then mount the remainder on the marker housing.

Step 3

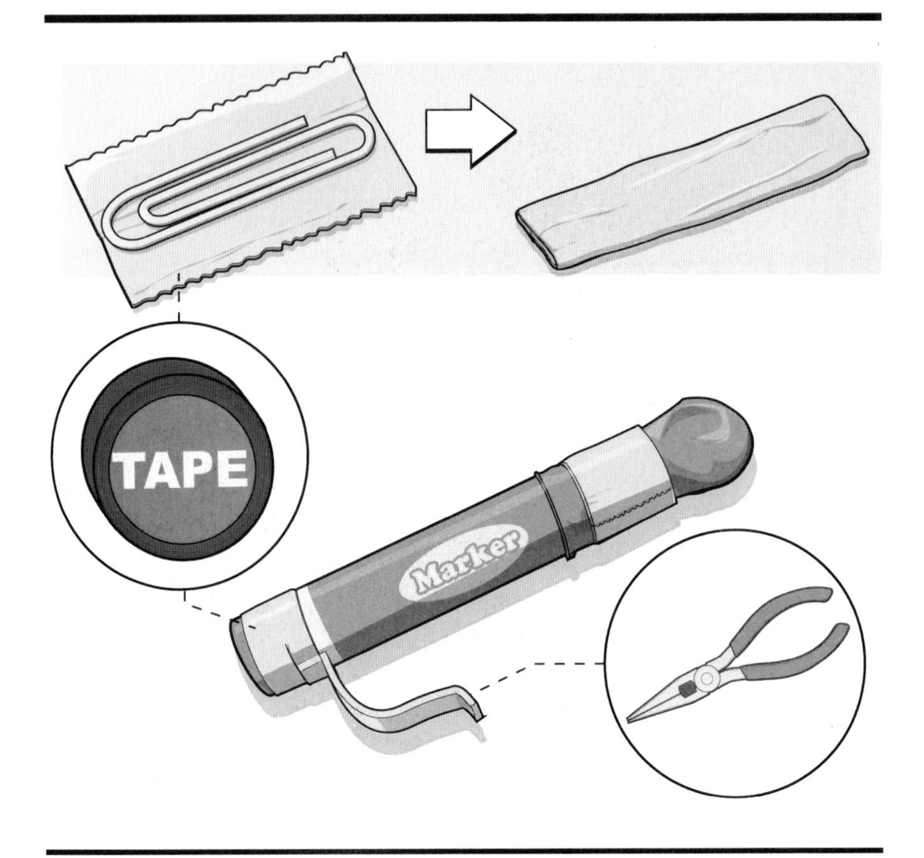

The trigger guard is constructed with a large paper clip and duct tape. Place the clip onto the sticky side of a slightly larger piece of duct tape. Securely wrap the clip so it's completely covered. Use pliers to slowly bend the clip to resemble a trigger guard as shown in the illustration. You should be making three bends in the clip, starting with the first bend in the center at 90 degrees.

Now affix the trigger guard onto the marker barrel as shown. The trigger guard should start approximately ½ inch from the tip of the barrel. Once in place, use duct tape to secure it.

Step 4

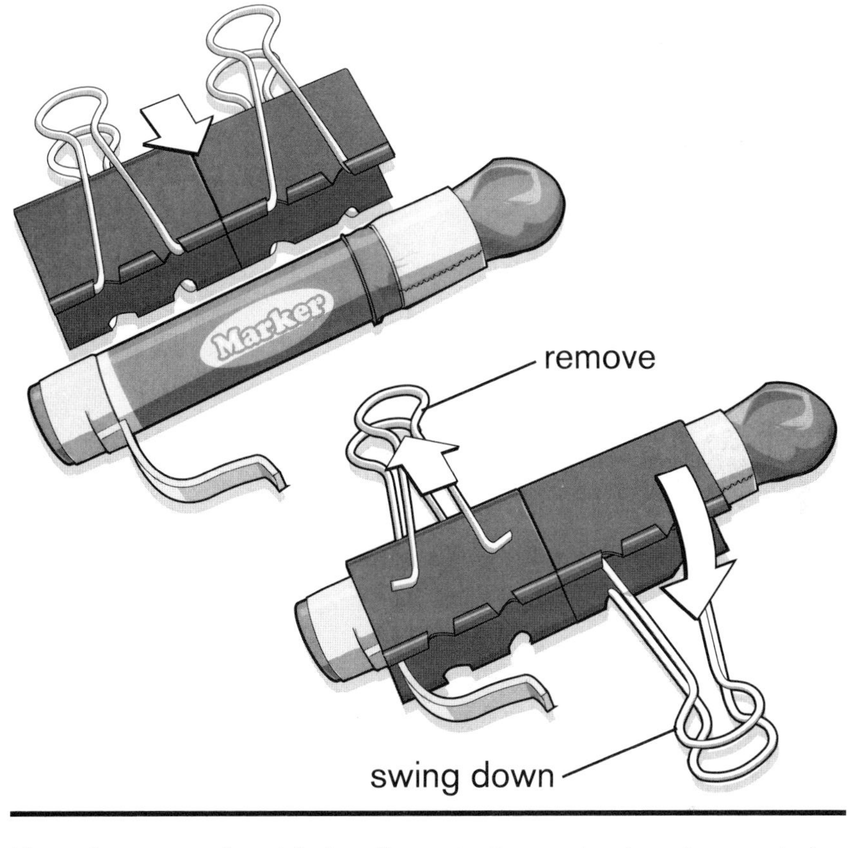

Now clip two medium binder clips onto the marker housing, centering them side by side in the middle. The clips will be positioned on the opposite side of the mounted trigger guard.

Remove the clips' front set of metal handles (opposite from the balloon), while flipping down the rear set of handles.

Step 5

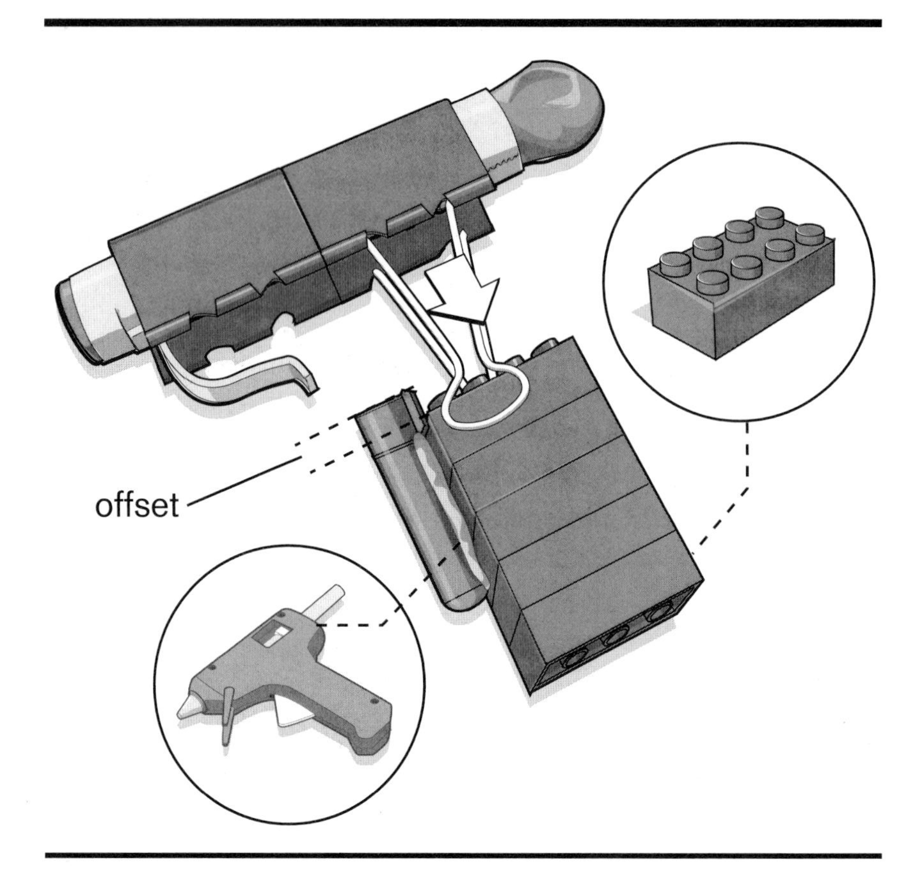

Start the construction of the Vest Pocket Mini's handle by assembling the 2-peg-by-4-peg building blocks into a five-block column.

Next, snap or cut off the clip of one plastic pen cap. Then hot glue or tape the modified cap to the center of the side of the block column. The pen cap should protrude approximately 1⁄4 inch off the top nub end.

Slide this custom handle between the two metal binder clips.

Step 6

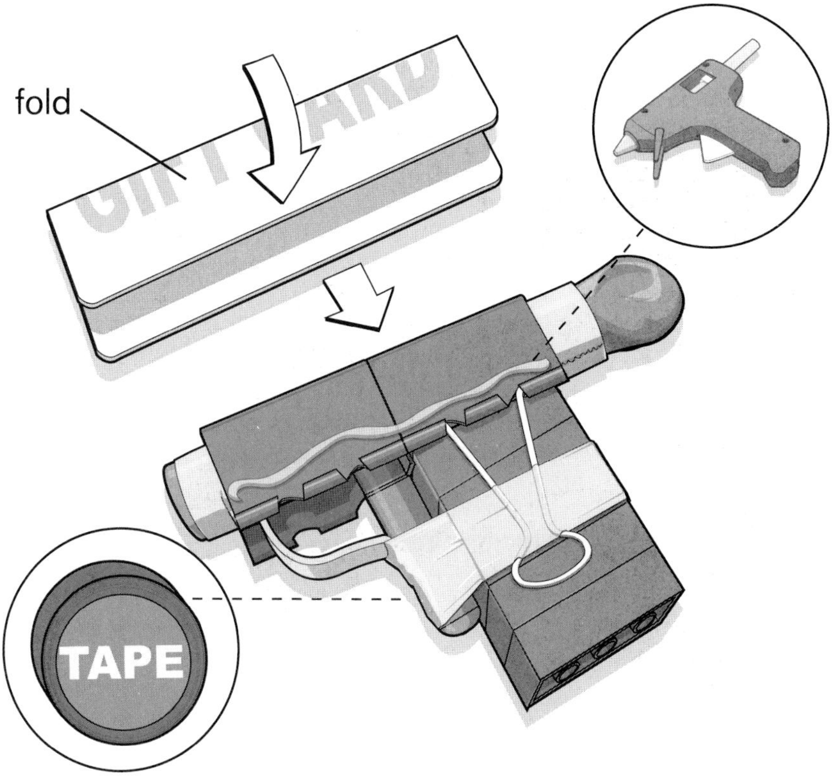

With the handle assembly between the two metal binder clip handles, duct tape it into place by wrapping the tape around the cap, blocks, and clip handles as shown.

To complete the junior pistol housing, fold the plastic gift card into three parts using two creases, the center being the width of the binder clips. Hot glue this molded card onto the binder clips to add styling.

Alternate Construction

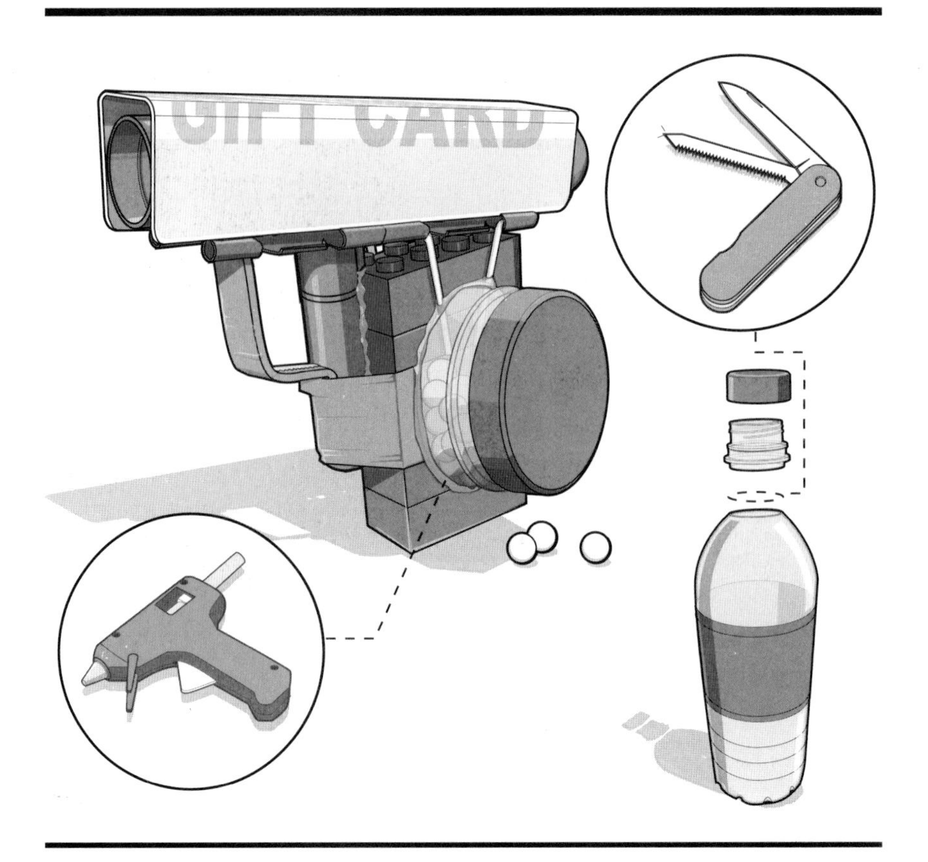

Because the Vest Pocket Mini is so compact, it lacks ammunition storage. This problem can quickly be solved with a side-mounted clip with removable cap.

Using a pocketknife blade, carefully saw or cut off the threaded neck of a soft drink bottle, similar to the firing mechanism for the Candy Glock 33 (see page 15).

Once you have removed the neck, hot glue it to the side of the Vest Pocket Mini as shown. Right-handed shooters should mount it to the left side of the handle, while left-handed shooters should mount to the right side. Once the neck is mounted and the glue has cooled, load the cap with candy ammo and seal it using the original bottle cap.

THE GOLDEN GUN

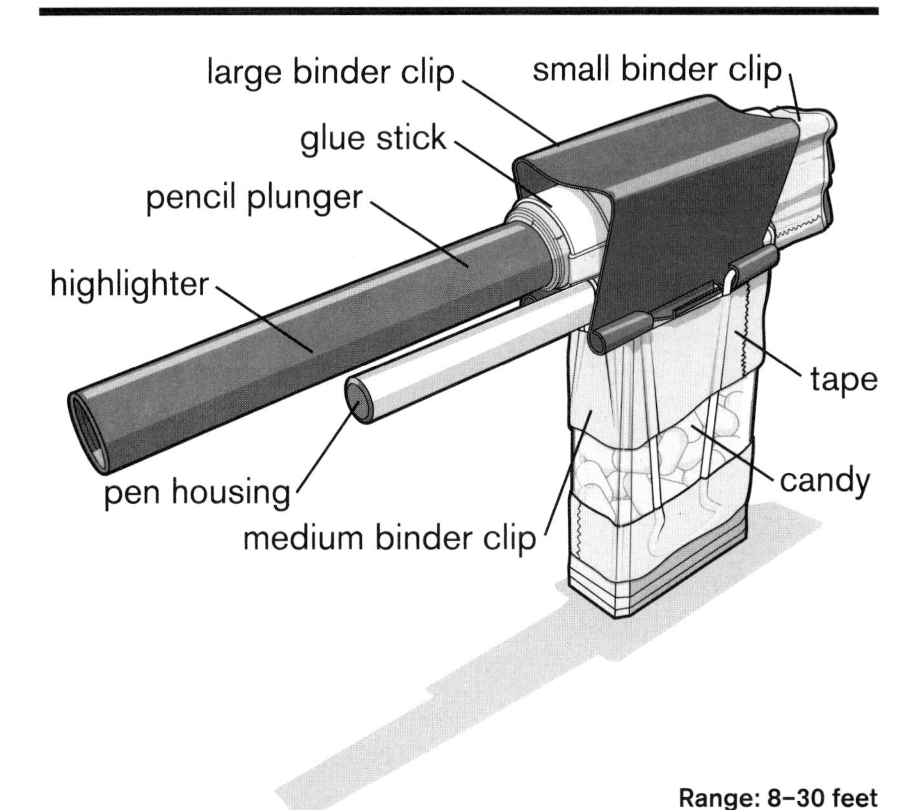

Range: 8–30 feet

The Golden Gun is one of the most popular villain weapons in spy history. Our reverse-engineered version is constructed out of everyday objects, not gold, but its marksmanship is still amazing at close range.

Supplies

1 inexpensive mechanical pencil
1 plastic ballpoint pen
1 glue stick (0.28 oz.)
1 medium highlighter
Duct tape
1 wide rubber band
1 medium binder clip (32 mm)
1 Tic Tac container
1 large binder clip (51 mm)
2 small binder clips (19 mm)

Tools

Safety glasses
Hobby knife
Pliers, wire cutters, or industrial scissors
Scissors

Ammo

1+ small, hard candies

Step 1

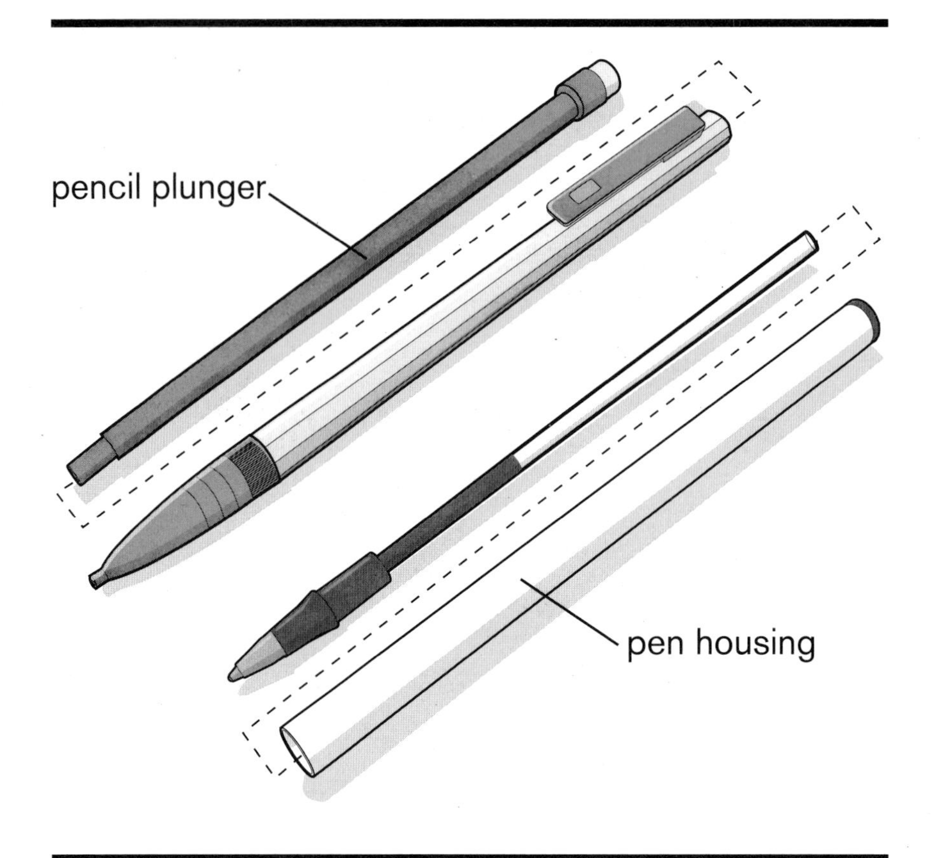

Dissect an inexpensive mechanical pencil using brute strength. First, pull out and save the pencil plunger. The outer housing of the pencil can be discarded or saved for a future MiniWeapons build, but you won't need it for this project.

Next, disassemble a plastic ballpoint pen by removing the tip and ink cartridge. You do not need to remove the rear pen-housing cap, however. Keep the pen housing and discard the rest of the pen's components.

Step 2

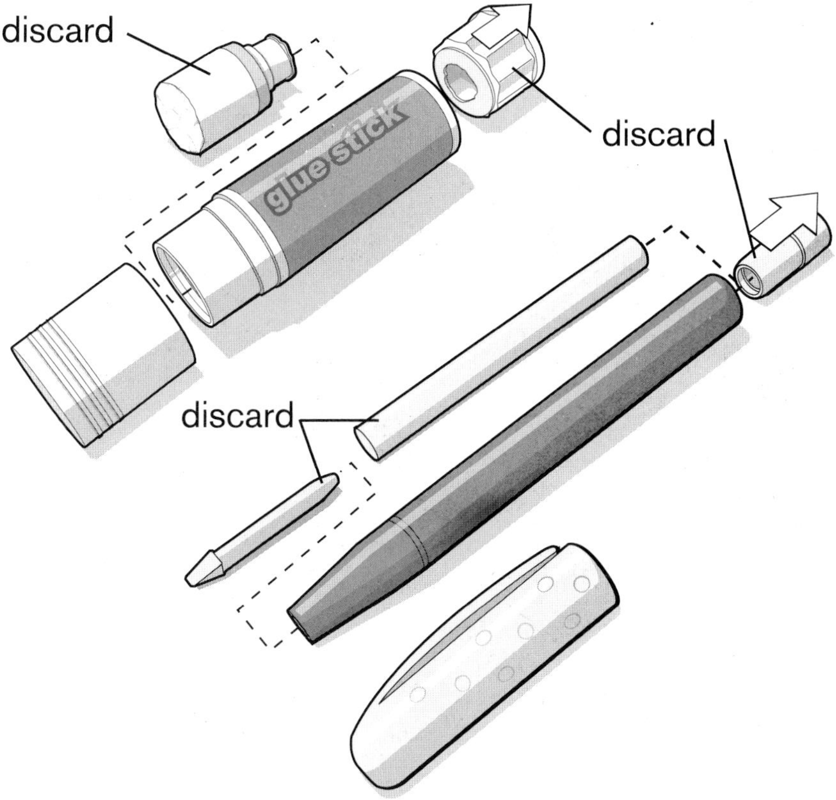

Grab a 0.28 oz glue stick and remove its cap, twist end, and glue insides, including the push rod. The twist end should just snap off, but use pliers if you find it difficult. Discard the glue and twist end; however, save the cap for the Q-pick Blowgun (page 111).

Next, you will dismantle a medium highlighter. Carefully remove the nib and inside ink canister. (If the housing contains ink residue, rinse it out and dry it.) Discard these components, except for the highlighter cap; the cap can be used for the Shark with Laser Beam (page 241).

Step 3

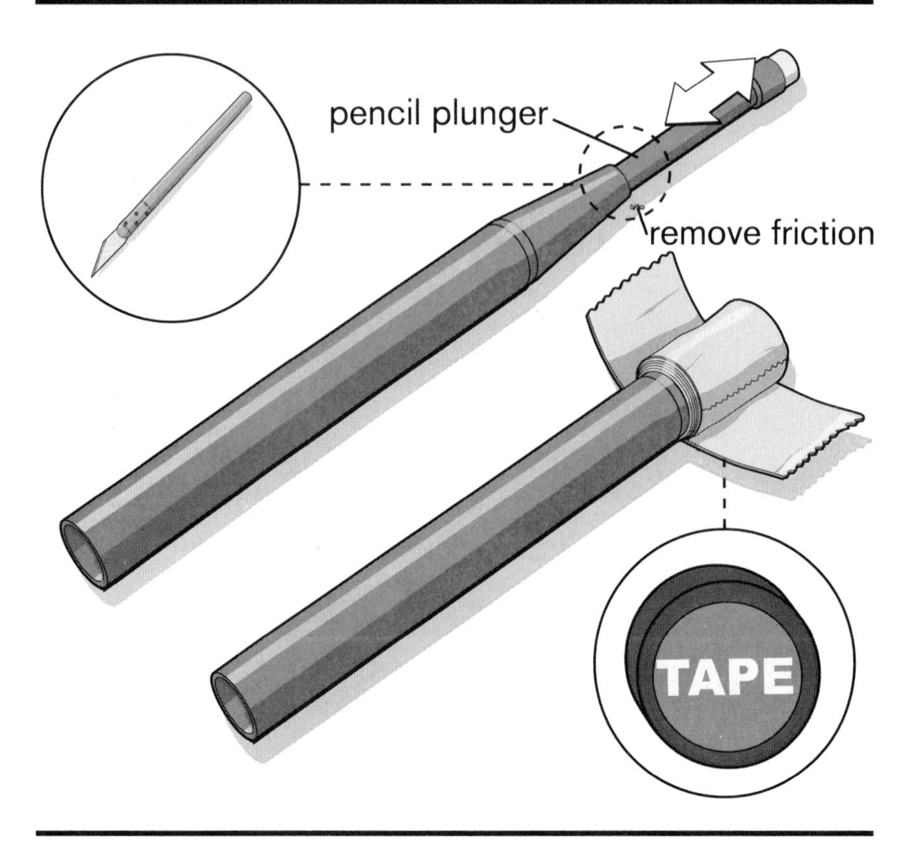

Using a hobby knife, increase the *inner* diameter of the narrow end of the highlighter housing. This can be done by slowly rotating the blade and carefully removing shavings or by shortening the housing by removing ¼ inch of material at the end. The ultimate goal is to have the pencil plunger slide easily in and out of the highlighter.

Next, use duct tape to increase the highlighter's *outer* diameter to match the inside diameter of the glue stick housing.

Step 4

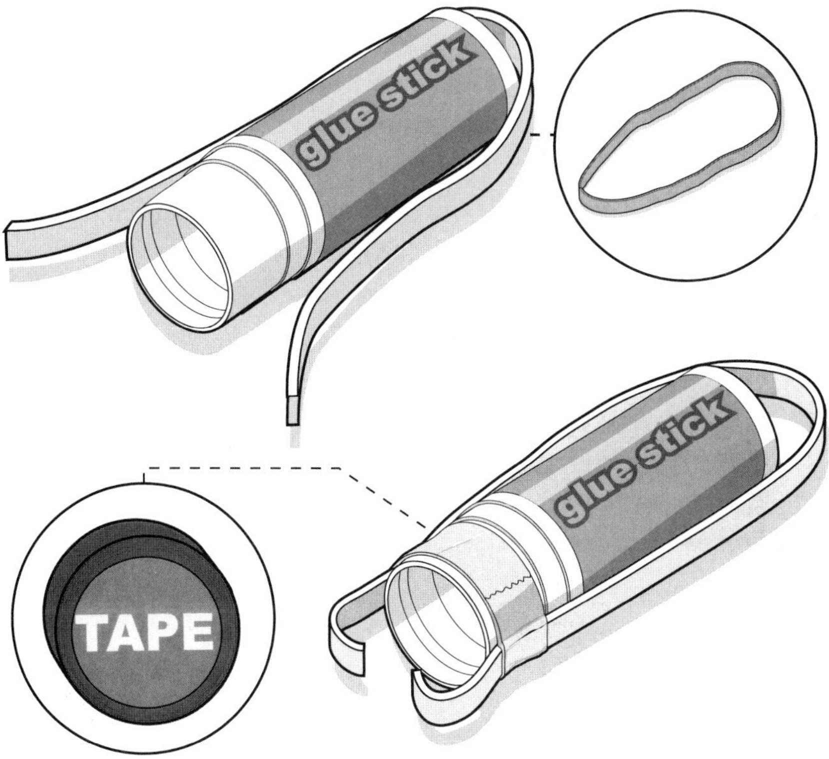

Using scissors, cut a wide rubber band to separate the loop. Then wrap the band around the bottom of the glue stick housing, centering it on the end of the tube as shown. Leave about ½ inch of slack—but not more—on the looped end before taping the rubber band.

The two loose rubber band ends will be tucked into the housing in the next step.

Step 5

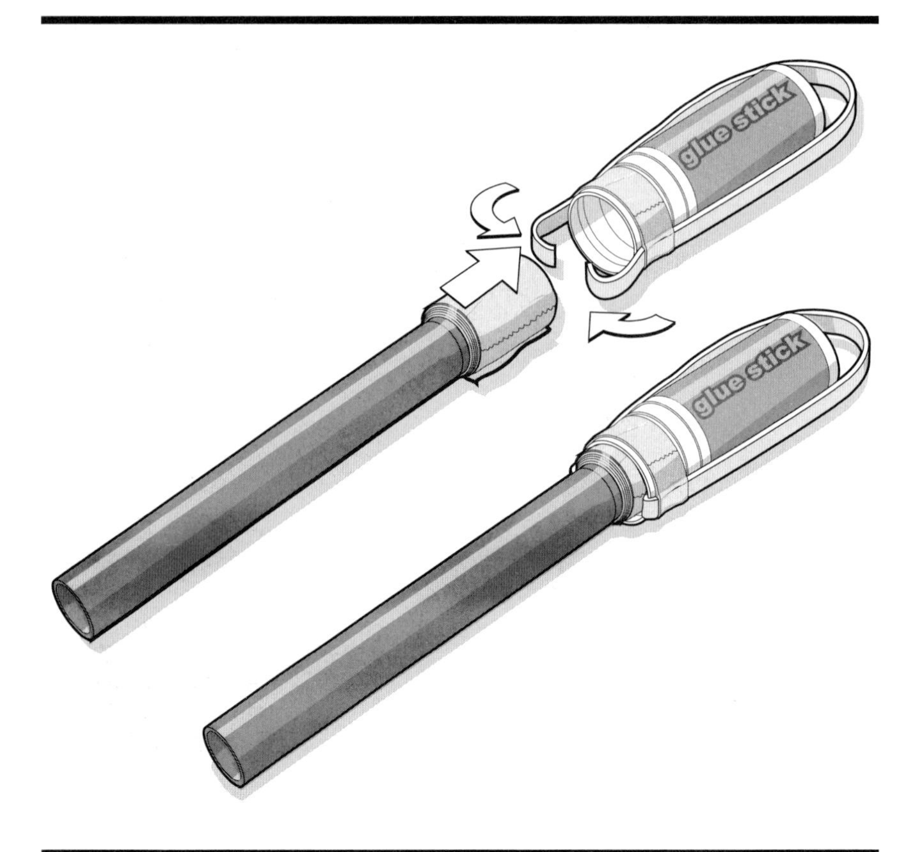

Slide the taped end of the highlighter into the glue stick housing. The fit should be snug, with the rubber band ends tucked into the housing for added strength. If the highlighter housing is loose, add additional duct tape around the highlighter to increase its outer diameter.

Step 6

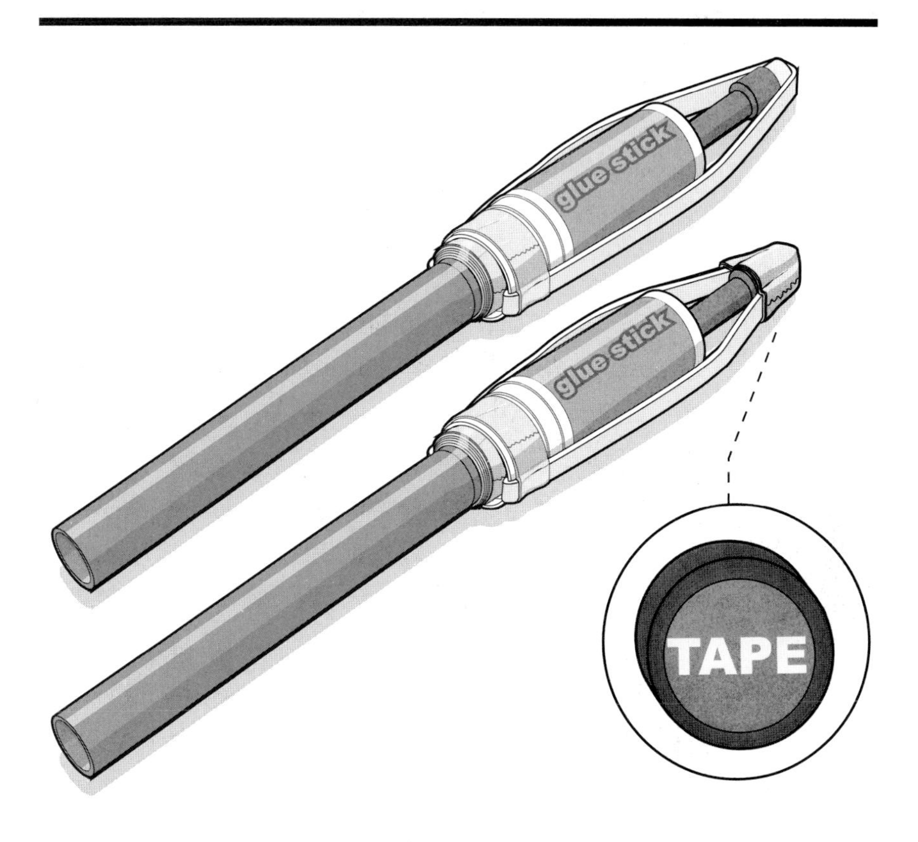

Now push the rubber band loop to the side and slide the pencil plunger (from step 1) into the twist end of the glue stick housing. Once in, wrap duct tape around the eraser end of the pencil and rubber band. You will add additional elements to this detail in a later step.

Step 7

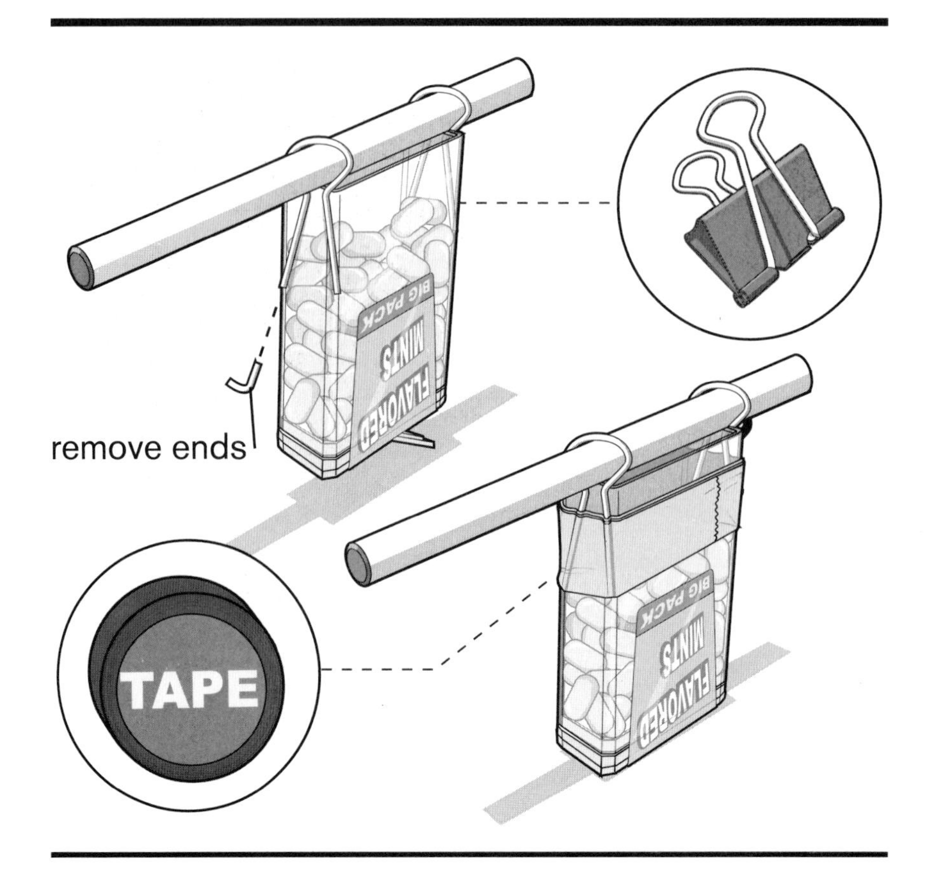

The trigger feature of the Golden Gun is very unique, a mechanism that is not found on any other MiniWeapons launcher. It has a spring-activated trigger, which allows the gunner to hold the launcher and fire it with one hand.

To begin assembly, use pliers, wire cutters, or industrial scissors to remove the two metal handles from one medium binder clip. These ends will poke your hand if you aren't careful. The metal handle loops from this size binder clip should be large enough to fit the pen housing through. Slide both metal handles onto the pen housing, then position the housing onto the bottom of the Tic Tac container, on the opposite side of the dispensing end. Tape both metal handles onto the container, as shown. The pen housing should slide back and forth with limited friction but not be loose.

Step 8

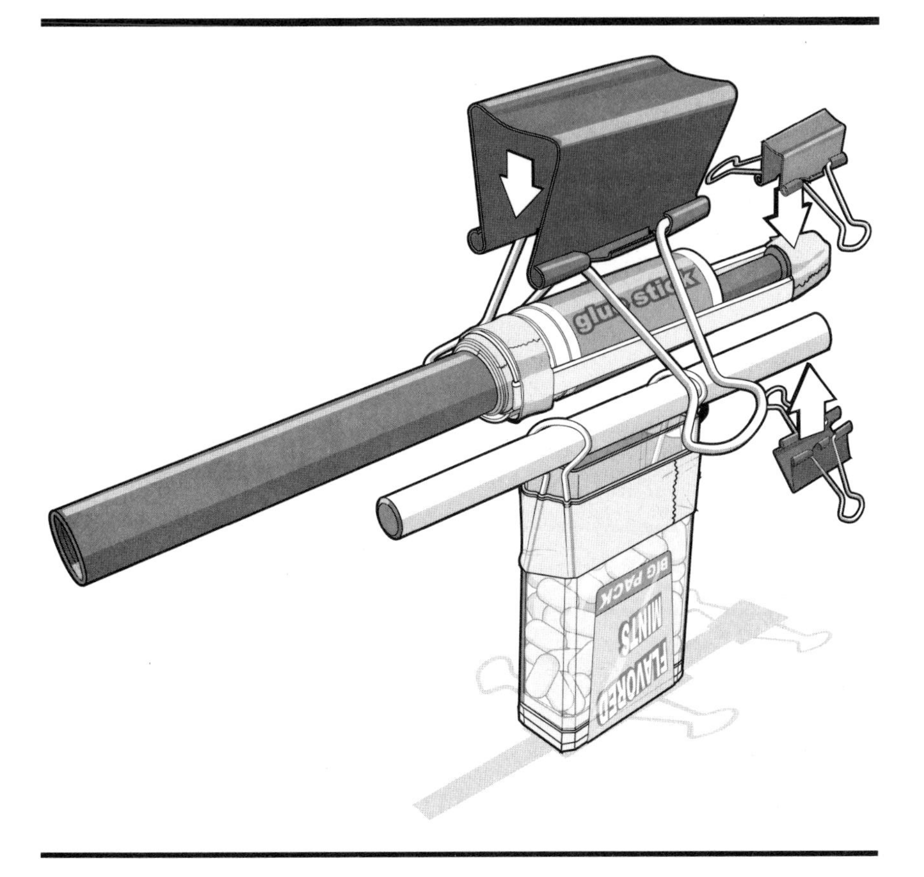

OK, agents, it's time assemble the Golden Gun using binder clips. Place the glue stick barrel assembly on top of the sliding pen housing and use a large binder clip to secure it into position.

Next, snap one small binder clip over the eraser end of the tapped plunger, then remove the metal handles. Mirror the same action on the pen housing from below. Once it's in place, remove the handles again.

Step 9

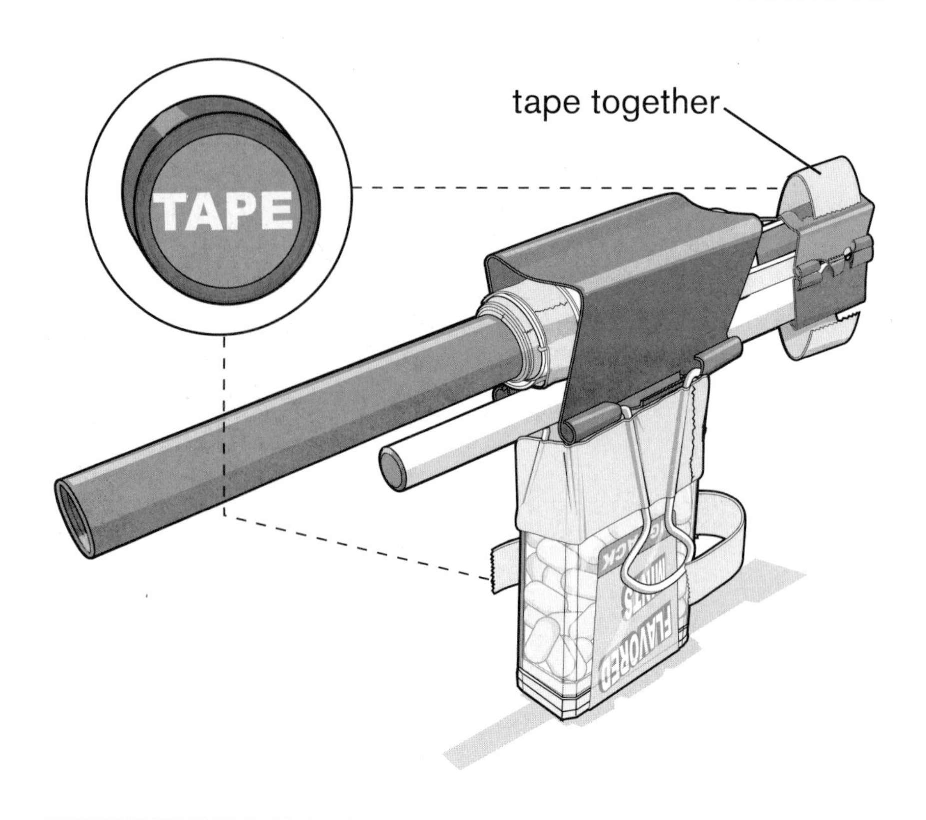

To finish the Golden Gun, use duct tape to secure the large binder clips and barrel assembly to the breath mint package.

Finally, tightly tape both small binder clips together to complete the trigger assembly. Once they are combined, load the barrel with a Tic Tac or other small, hard candy, use your trigger finger to push the elastic plunger back, then quickly release it to fire!

Troubleshooting distance: To increase the Golden Gun's firing distance, reposition the rubber band to maximize the elastic tension. If built correctly, this launcher will fire ammo a great distance.

Troubleshooting misfires: Just before firing, when the trigger (and plunger) is pulled back, tilt the barrel backward to allow the candy to rest on the plunger, then release the trigger.

BINDER BERETTA 92

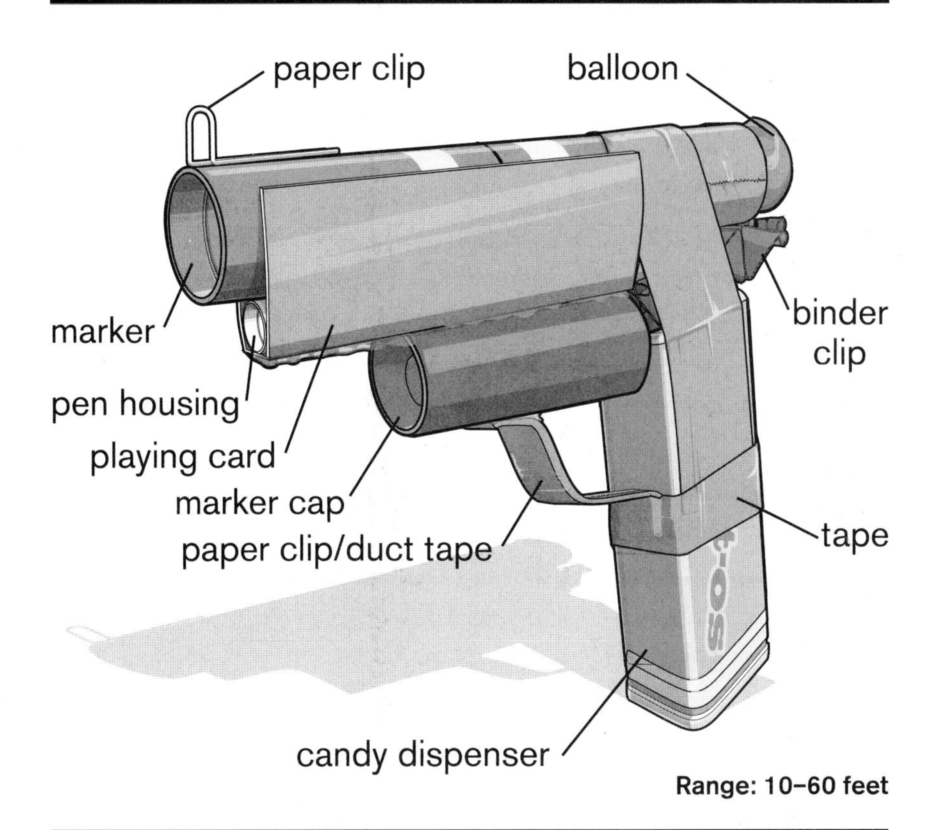

Range: 10–60 feet

The Binder Beretta 92's single-action, long-barrel design is counter-weighted by a large candy dispenser. The generic design encourages mission-specific upgrades, such as a front-mounted flashlight or laser pointer.

Supplies

3 small binder clips (19 mm)
1 plastic ballpoint pen
2 standard art markers
1 balloon
Duct tape
1 plastic mint gum container
1 playing card
1 small paper clip
1 large paper clip

Tools

Safety glasses
Hot glue gun
Hobby knife
Pliers
Scissors

Ammo

1+ small, hard candies

Step 1

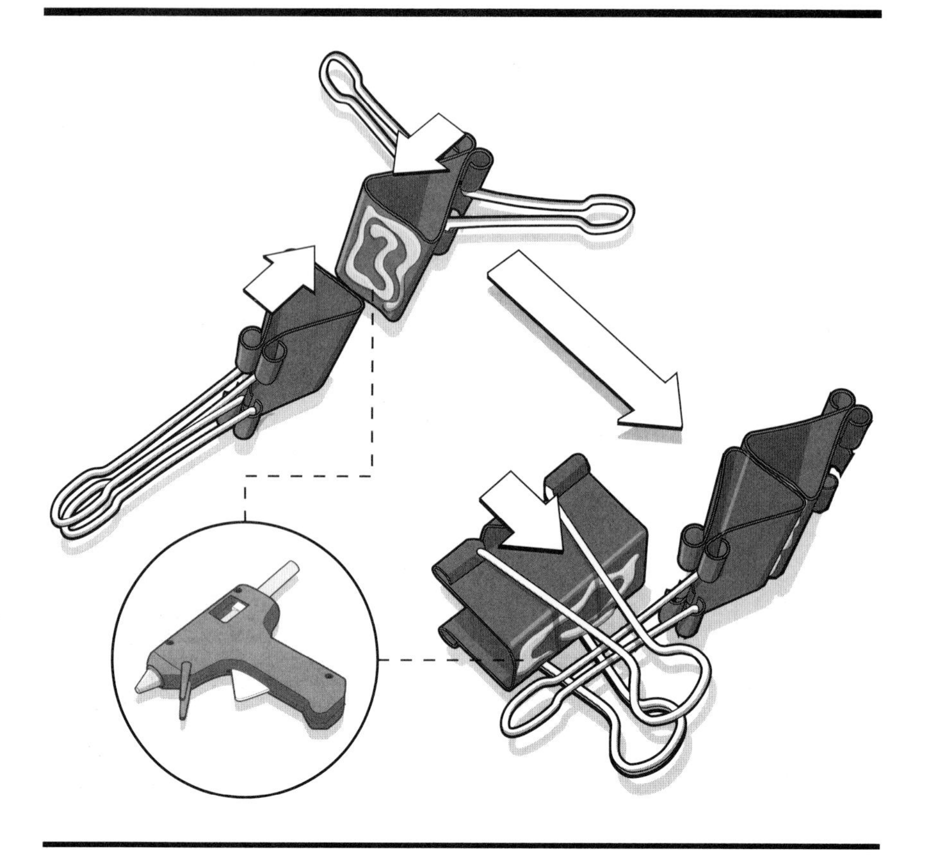

Place hot glue on the bottoms of two small binder clips, then carefully push them together and hold them until the glue cools. Then, remove three of the four metal handles as shown in the bottom graphic.

Next, hot glue a third small binder clip to the top of the remaining metal bracket.

Step 2

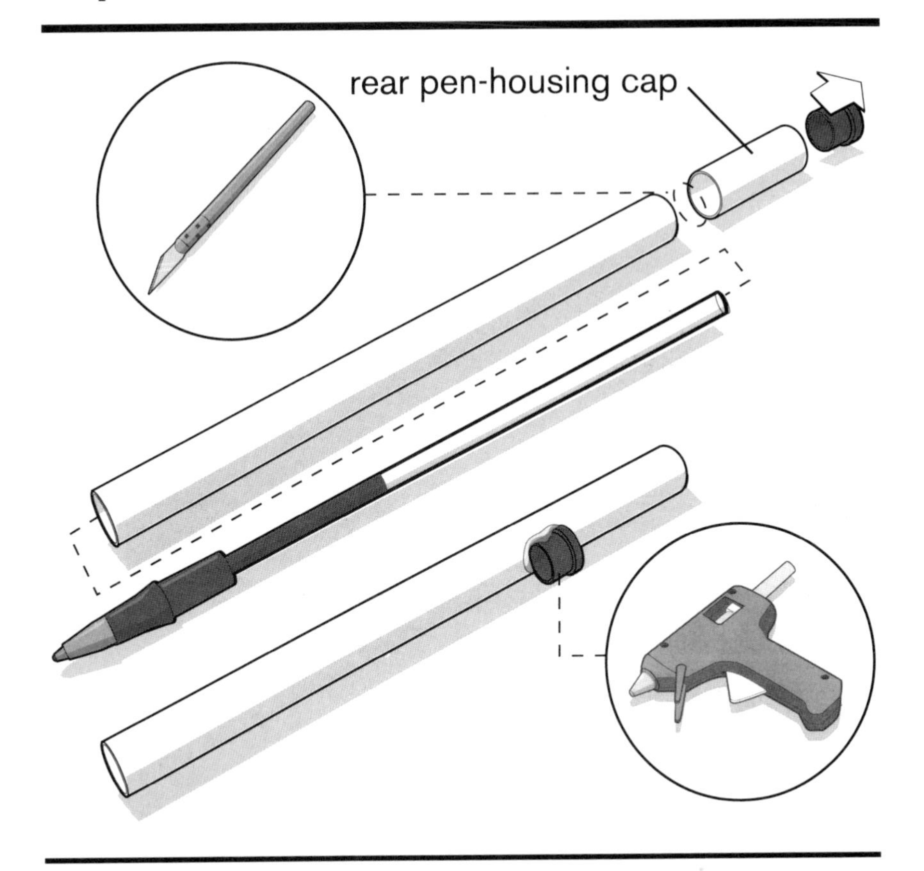

Next, disassemble a plastic ballpoint pen by removing the tip, ink cartridge, and rear pen-housing cap. If you need a tool to dislodge the rear pen-housing cap, use a hobby knife (carefully!) or a small set of pliers.

Now, use a hobby knife to shorten the pen housing length by 1 inch. The small removed section can be discarded. Then, using a hot glue gun, reattach the pen-housing cap to the pen housing, approximately ¾ inch from the end.

Step 3

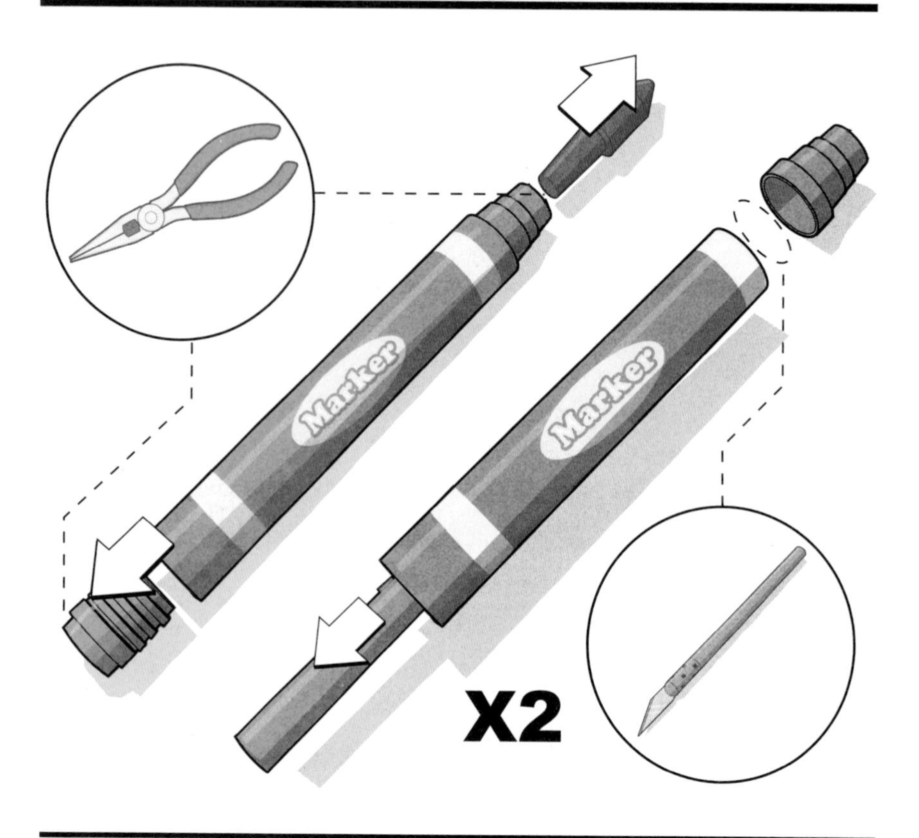

The length of this launcher will require you to modify two standard art markers. Again, if possible, a recycled plastic marker is preferred because of the material softness, but any marker will work. Using pliers, remove the end cap by slightly rotating it from the housing. Use pliers to remove the marker nib by sliding it out the tip of the marker. Discard this nib to avoid a mess.

Using a hobby knife, pierce the housing below the plastic point on the marker housing and slowly rotate the marker; this will make cutting it safer. Once this end has been completely removed, the internal ink cartridge should slide out. Discard this cartridge. Repeat this for both markers. (If the housing contains ink residue, rinse it out and dry it.)

Make sure that the housing pathway is unobstructed by any plastic fragments that might have broken off during cutting. Save one of the marker caps; the other can be discarded.

Step 4

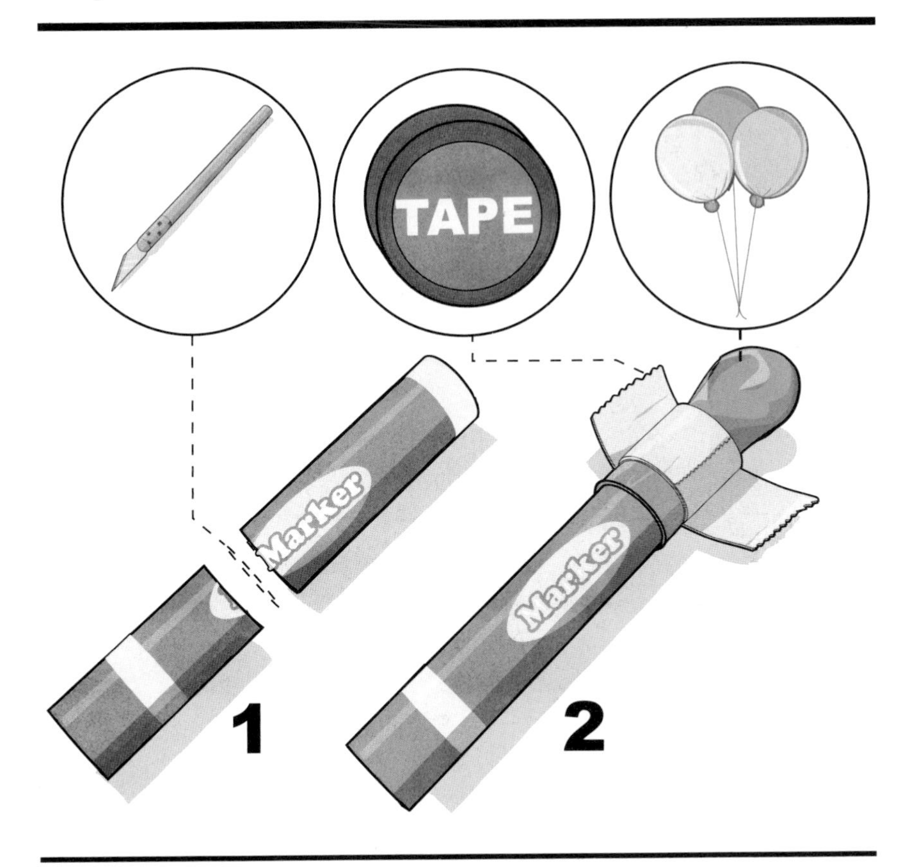

Measure approximately 2 inches from the end of one marker and slowly cut it off with a hobby knife.

Attach a small balloon to one end of the second marker housing. The majority of the balloon should cover the housing and roughly ½ to ¾ inch should overhang the end. Tape the balloon securely into position. Test the tape restraint a few times, and add more tape if needed. It is important that you do this before you begin construction of the rest of the gun.

If the balloon is too large, use scissors to remove some of the balloon material from the mouthpiece end, then mount the remainder on the marker housing.

Step 5

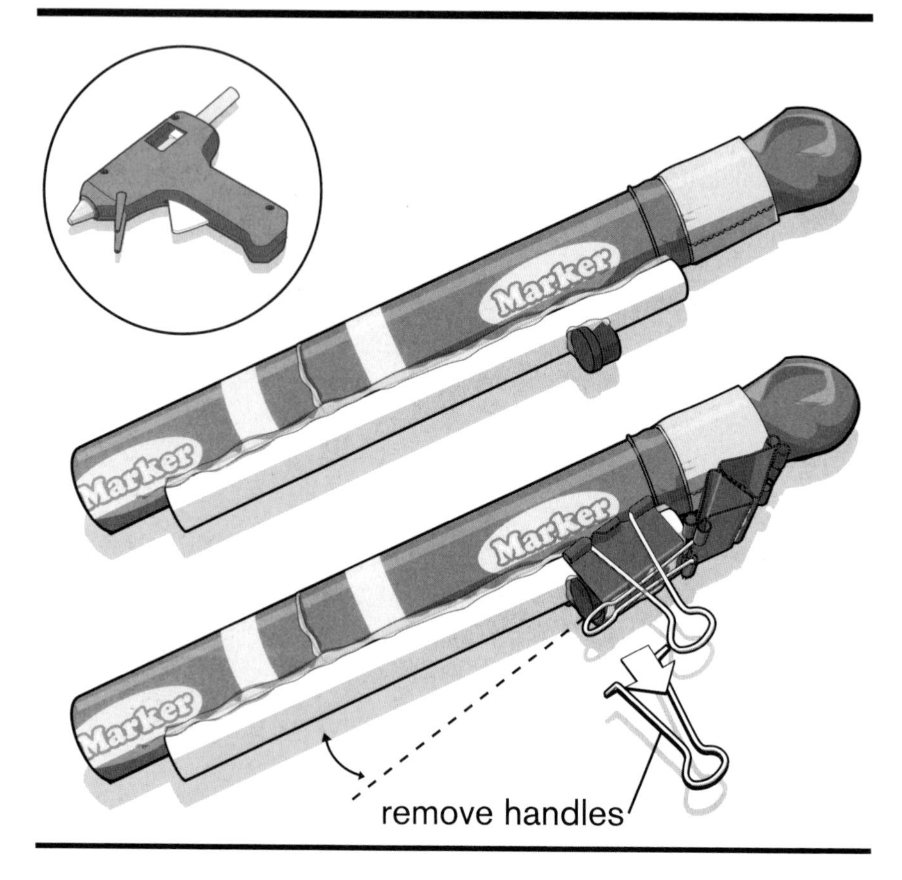

Use tape or a hot glue gun to join both marker housings to the shortened pen housing, creating one long cylinder. The pen housing should be set back approximately 1/4 inch from the front of the barrel as shown.

Next, clip the binder clip assembly onto the pen housing and pen housing end cap. Once in place, it should be on a slight angle (see illustration); this is preferred. When you are happy with the position, remove the two metal handles, but not the glued handle.

Step 6

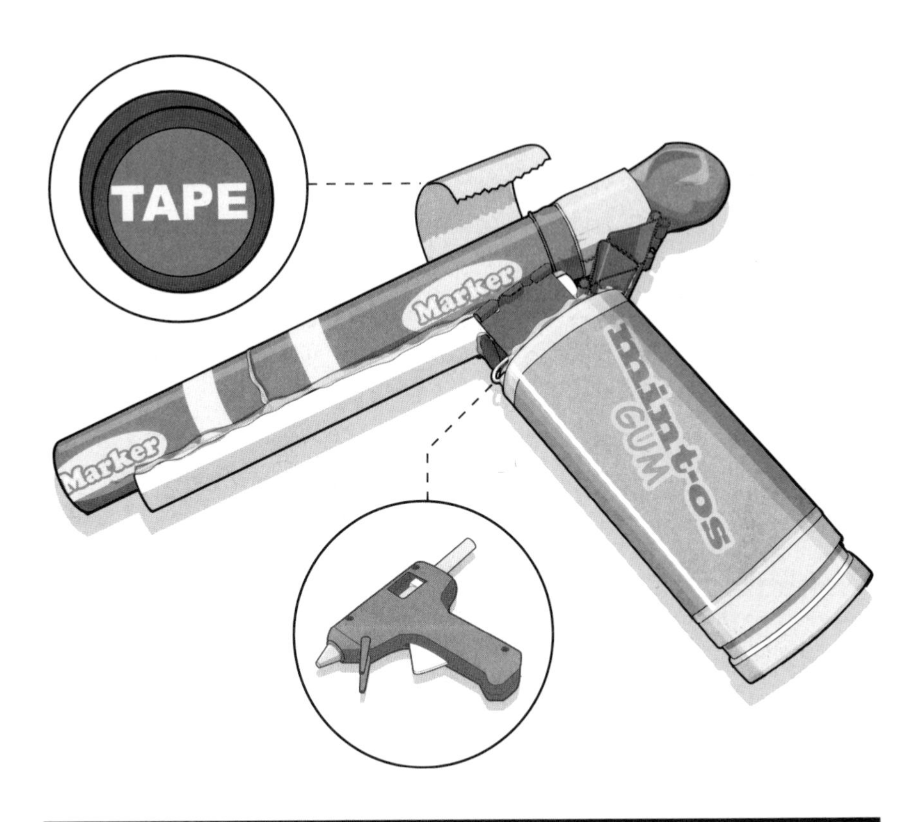

For the Binder Beretta 92 handle, we suggest a plastic mint gum container. If this is not available, you might find a substitute health or beauty package that is small and grip-like, such as a deodorant dispenser. Glue the container onto the binder clip assembly. After the glue has cooled, add additional tape to strengthen the connection.

Step 7

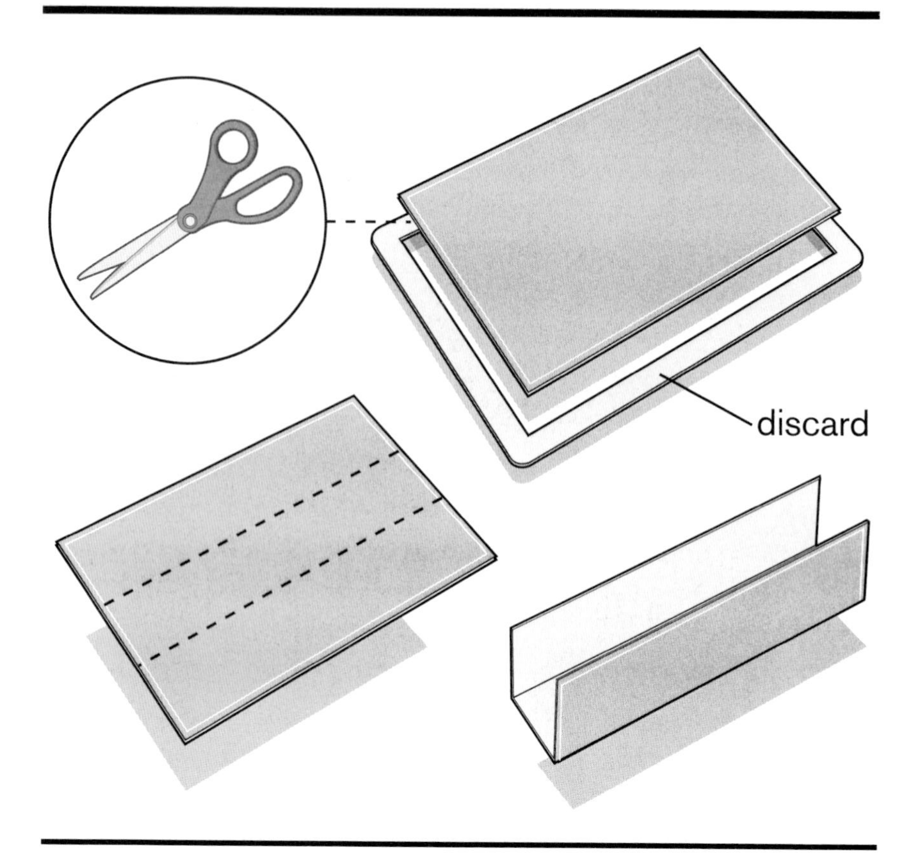

Now use a playing card to complete the industrial styling of this barrel.

First, use scissors to reduce the card's size. Cut around the trim of the playing card, removing approximately 3⁄16 inch, or just follow the printed border (if it's about the same width).

Next, fold the card into three sections, using two crease lines. The center section should be similar to the width of the pen housing.

Step 8

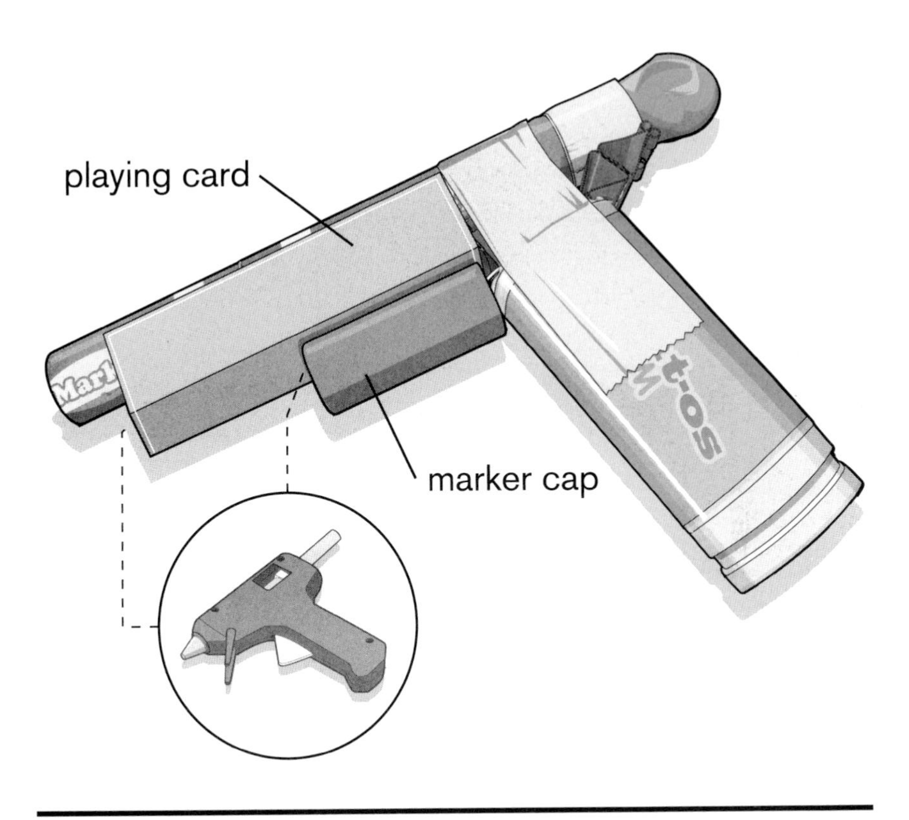

Glue the modified playing card underneath the pen housing as illustrated. Cut and adjust it to fit correctly if needed.

Next, glue one of the saved marker caps to the bottom of the playing card, pressed against the candy container as shown.

Step 9

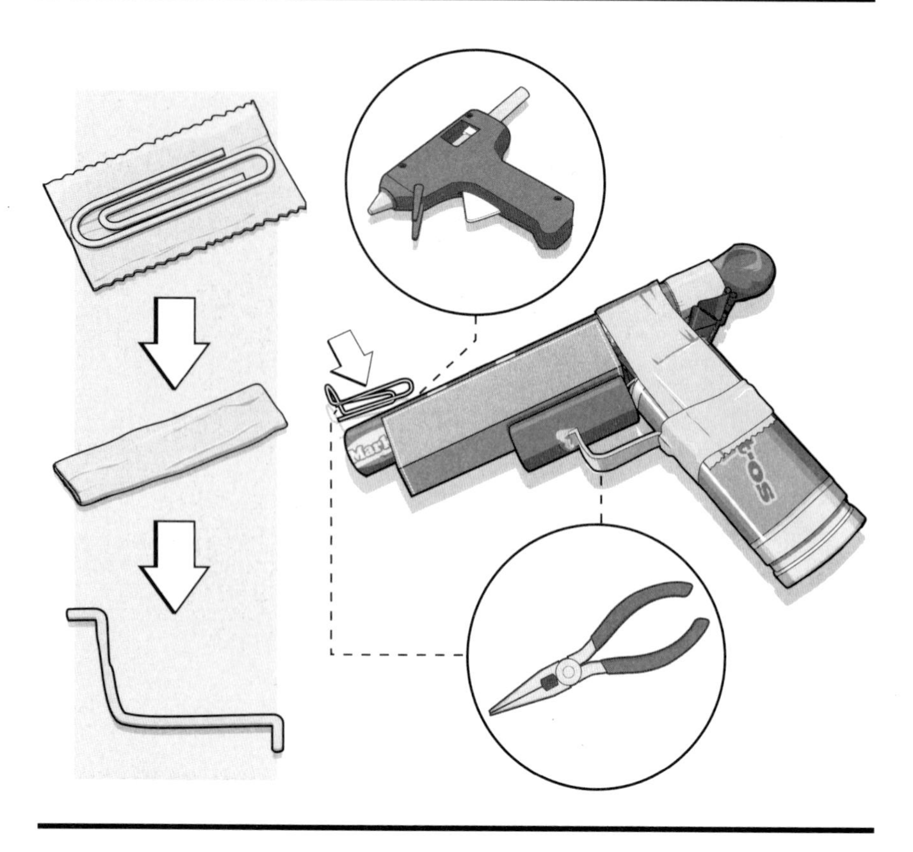

Use one small and one large paper clip for the final step. The small paper clip will be a simple barrel sight. Create it by bending one end of the clip at a 90-degree angle as shown. Use hot glue or tape to secure it to the end of the barrel.

The large paper clip is used for the trigger guard detail. Place the clip onto the sticky side of a slightly larger piece of duct tape. Securely wrap the clip so it's completely covered. Next, use pliers to slowly bend the clip to resemble a trigger guard as shown in the illustration. You should be making three 90-degree bends in the clip, starting with the first bend in the center. Once your trigger guard resembles the one in the illustration, hot glue or tape it to the launcher.

Before you suit up for your mission, load the barrel with some test ammo, such as a Tic Tac or other small, hard candy. Once the ammo slides into the balloon, locate it with your fingers and pull back. Find your (nonhuman) target and release!

44 MARKER MAGNUM

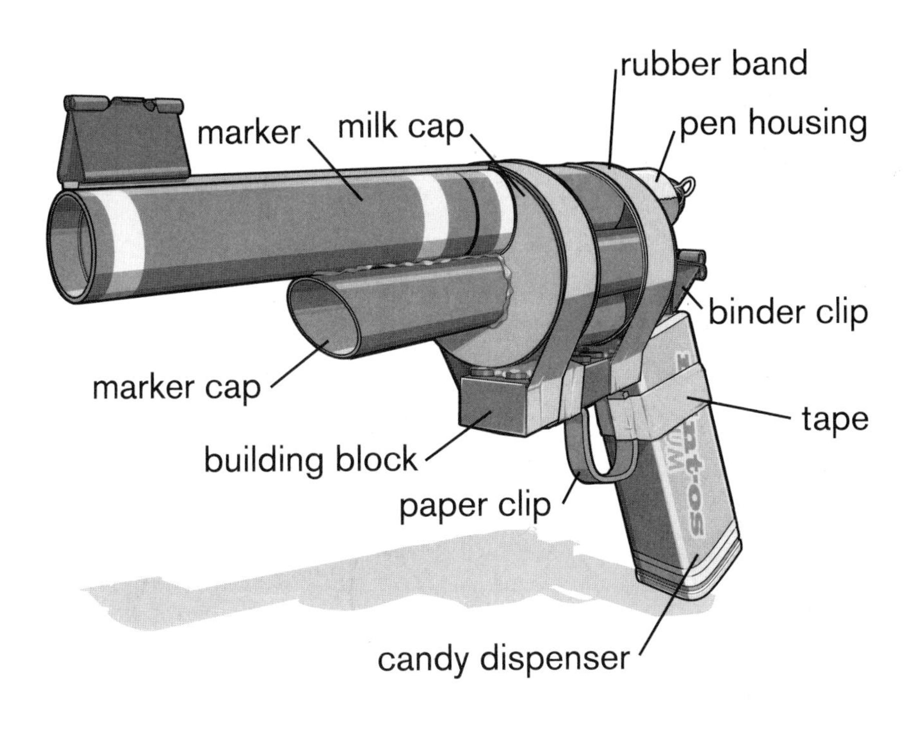

Range: 10–30 feet

Once called "the most powerful MiniWeapon in the world," the 44 Marker Magnum features a hefty cylindrical construction and an elastic pullback trigger. It is capable of firing small candy ammo at a high velocity and storing additional reloads in the grip handle.

Supplies

4 standard art markers
2 plastic milk jug caps
1 plastic ballpoint pen
1 small binder clip (19 mm)
2 rubber bands
Duct tape
1 craft stick
1 medium binder clip (32 mm)
2 large paper clips
1 2-peg-by-6-peg building block
1 plastic mint gum container

Tools

Safety glasses
Hobby knife
Pliers
Hot glue gun

Ammo

1+ small, hard candies

Step 1

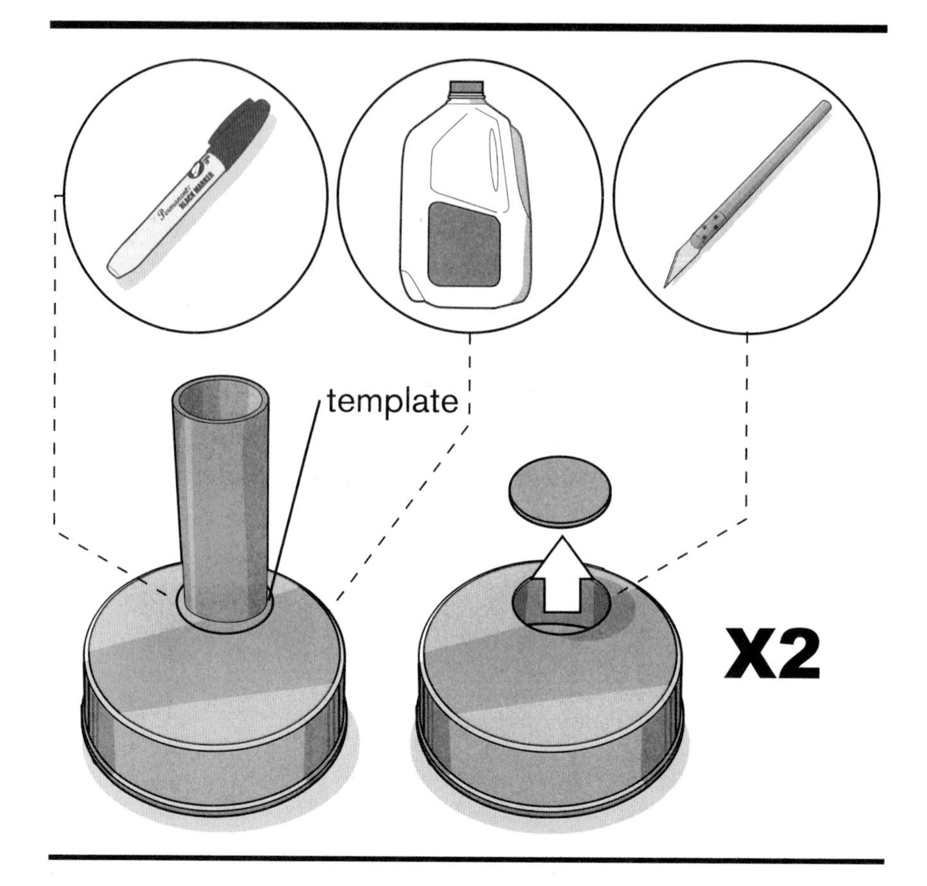

The revolver cylinder is made up of three marker caps sandwiched between two plastic milk jug caps. The marker barrel passes through the revolver cylinder, but before you get to that step you'll have to cut two small holes to allow for that.

First, use one marker cap as a template and trace the cap's diameter on top of two plastic milk jug caps, near the edge as shown. Then, carefully use a hobby knife to cut out the two circles. The circles do not have to be perfect.

Step 2

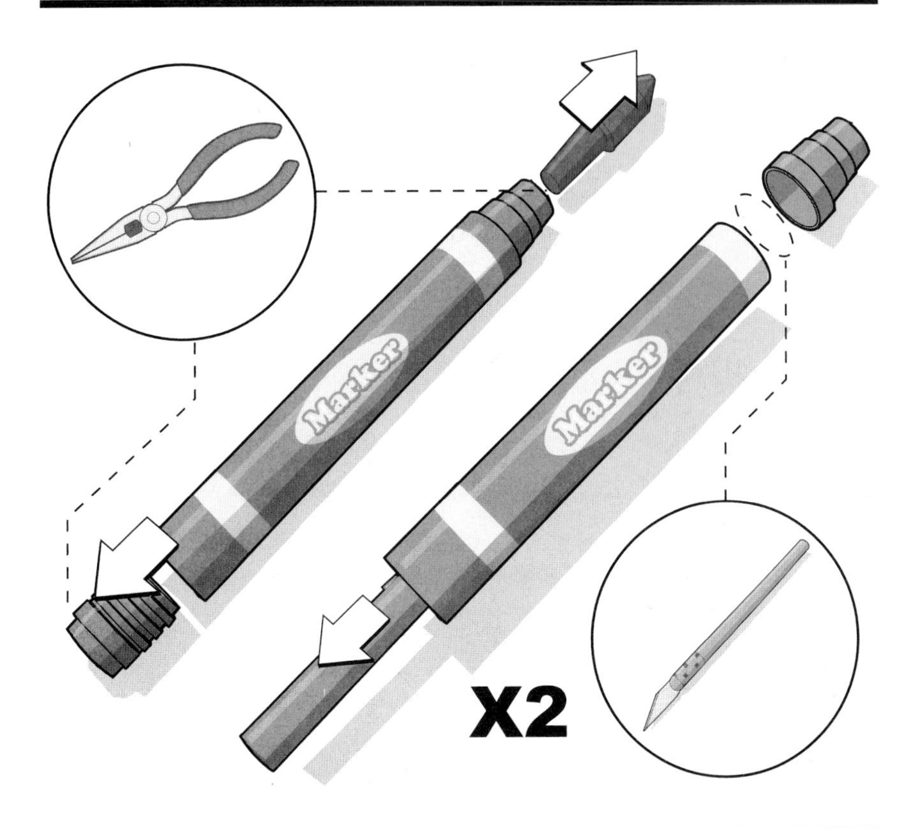

To construct the powerful 44 Marker Magnum barrel, you will use the length of two standard art markers, each with a ½-inch diameter. If possible, use art markers manufactured with recycled plastic; the material is thinner, making it easier to cut. Using pliers, remove the end cap by slightly rotating it from the housing. Then, use the pliers to remove the marker nib by sliding it out the tip of each marker. Discard the nibs to avoid a mess. Save one of the marker caps; the other can be discarded.

Using a hobby knife, pierce the housing below the plastic point on the marker housing and slowly rotate the marker. This will make cutting it safer. Once the end has been completely removed, the internal ink cartridge should slide out. Discard this cartridge. Repeat this for both markers. (If the housing contains ink residue, rinse it out and dry it.)

The barrel housing pathway should be unobstructed by any plastic fragments that might have broken off during cutting.

Step 3

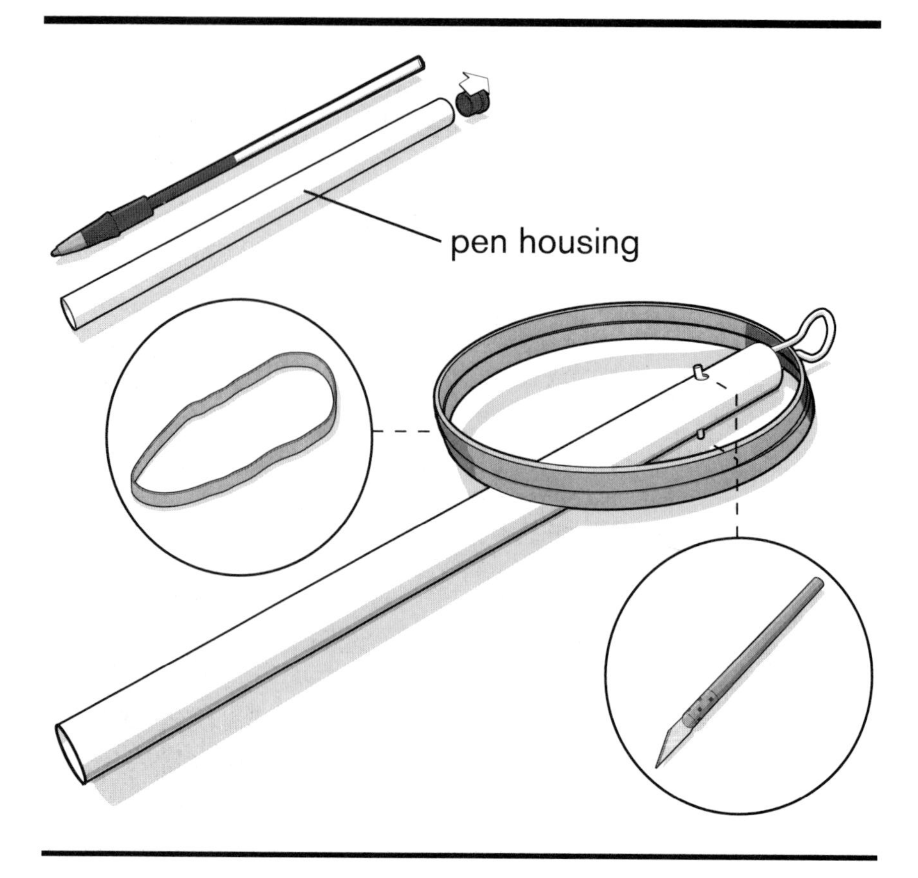

Disassemble a plastic ballpoint pen. You may need a hobby knife or pliers when removing the rear pen cap. The hollowed-out pen housing will be used for your 44 Marker Magnum's elastic trigger. Discard the other pen contents.

Next, use a hobby knife to make two small holes across from one another in the pen housing, about ½ inch from one end. These holes will be very small, just large enough to accommodate the metal handle wire diameter of a small binder clip.

Remove one metal handle from a small binder clip, then slide two rubber bands in between the rods. Once in place, slide the binder clip metal handles into the pen housing until they "click" into the two custom holes you created, popping out from the inside. Adjust the holes and length if needed.

Step 4

To prevent discomfort and add strength, wrap a small amount of duct tape around the end of the pen housing to cover the protruding metal handle tips.

Next, slide your 44 Marker Magnum's elastic trigger into one of the hollowed-out marker housings. Once in place, tightly tape the rubber bands to the marker housing. Pull back the trigger a few times to confirm the tape's holding power, and add tape if necessary.

Step 5

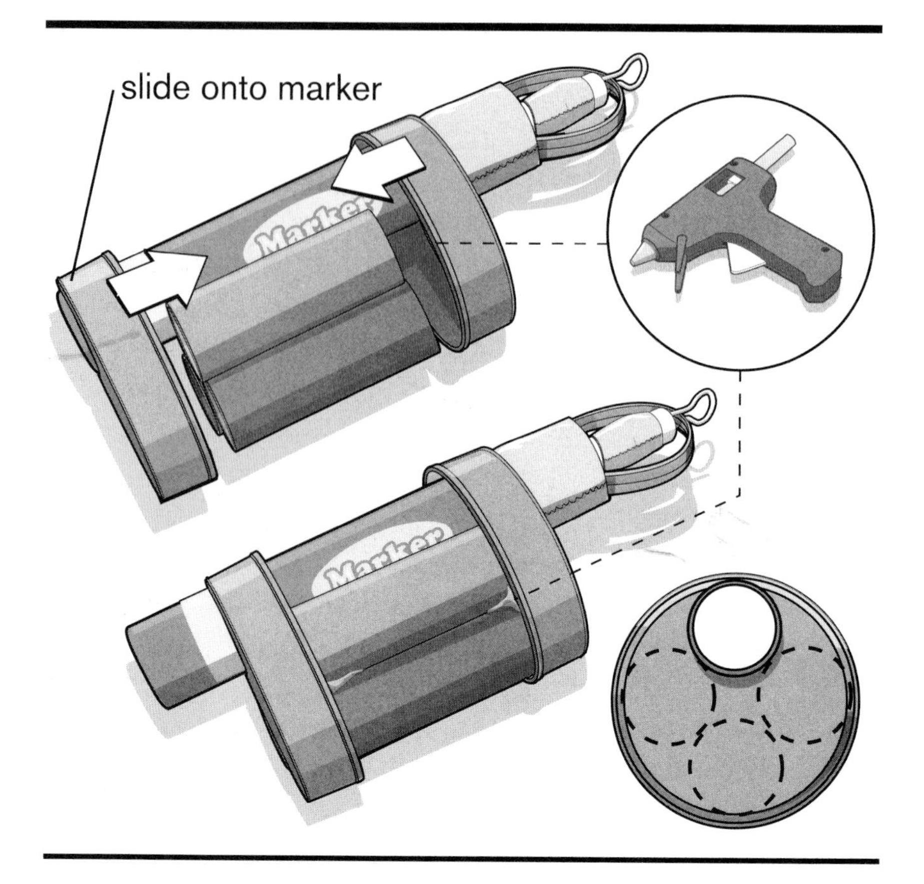

Now slide the marker housing with attached trigger mechanism through the two modified plastic milk jug caps, with the flush ends of the caps to the ends.

Place three marker caps in between the caps for a snug fit. Use hot glue to secure the whole assembly. The marker barrel should extend at least ¼ inch from the cap; ½ inch is ideal. Reference the illustration above for proper spacing and placement.

Step 6

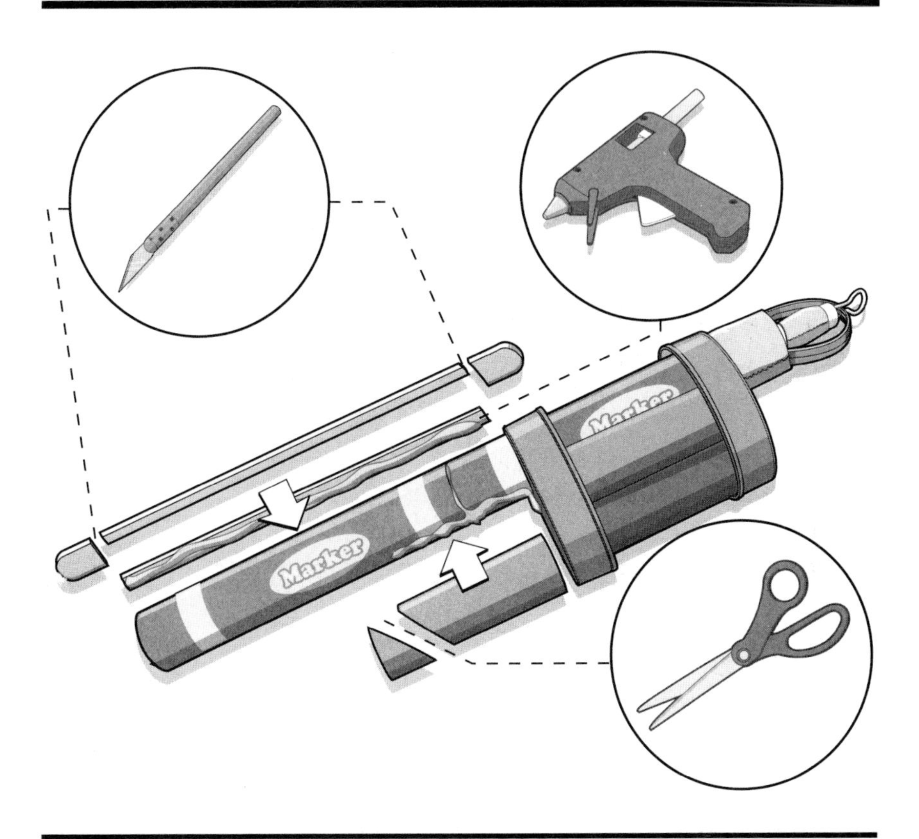

To extend the barrel, carefully hot glue the second hollowed-out marker housing to the end of the first marker housing. It is important that you limit the amount of hot glue inside the barrel because excess glue might interfere with the elastic trigger mechanism.

To add barrel support and styling, hot glue a single marker cap to the bottom of the barrel assembly as shown. You may want to use the hobby knife to add a slight angle to this marker cap to make it more realistic.

Additional barrel support comes from a cut-down craft stick mounted with hot glue to the top. Determine the length you need by holding it up to the maximum distance of the barrel. Then cut the craft stick's length and width to fit the top of the barrel. You may want to use a metal straightedge when cutting the stick's width.

Step 7

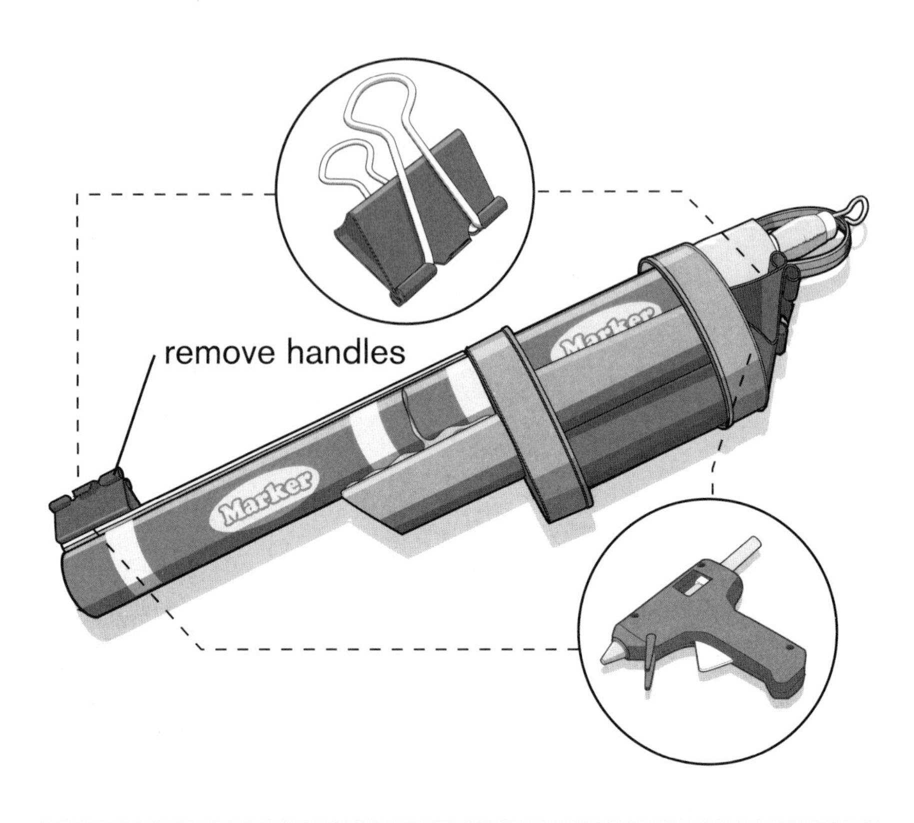

Remove the remaining metal handle from the small binder clip and both handles from a medium binder clip. Now, following the diagram, use hot glue to mount the clips to the 44 Marker Magnum. The small binder clip on the end of the barrel creates a serious sight for a serious gun. The medium binder clip should be glued to the back of the milk jug cap using plenty of hot glue; this will be the handle mount.

Step 8

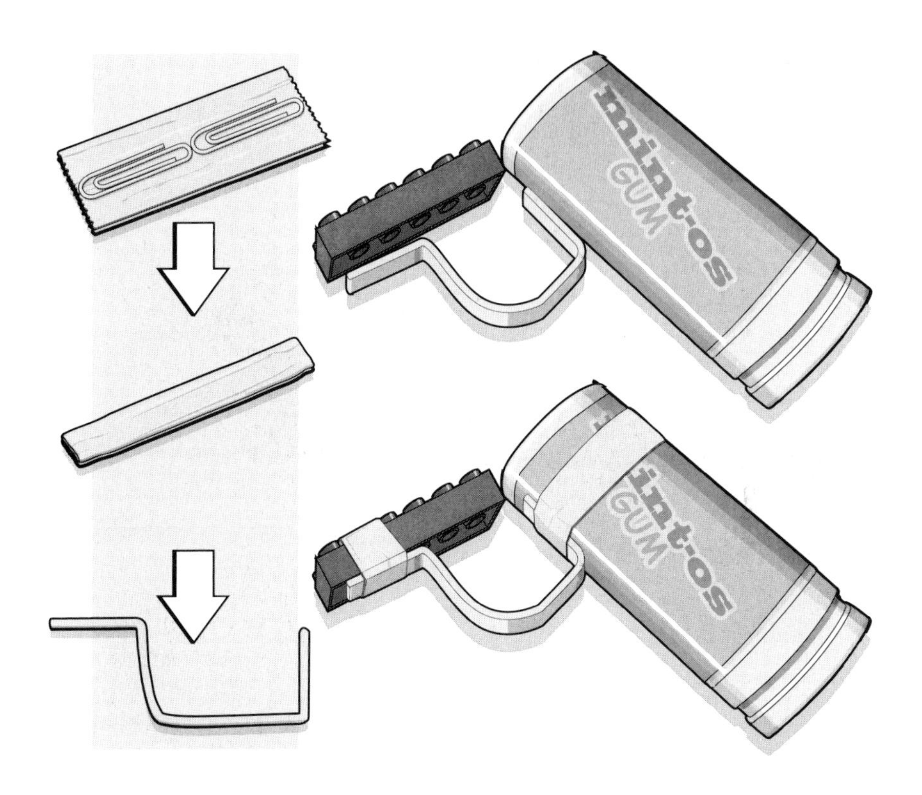

Now construct the trigger guard and grip detail. Use two large paper clips for the internal framework. Place both clips end to end onto the sticky side of a slightly larger piece of duct tape, as shown. Securely wrap the clips so they are completely covered with tape. Then use pliers to slowly bend the clips to resemble a trigger guard as shown in the bottom left illustration. You should be making three 90-degree bends in the clip, as shown.

Tape or hot glue the top, flat side of the trigger guard assembly onto a 2-peg-by-6-peg (or equivalent) plastic building block.

The rear of the paper clip trigger guard will be attached to the side of a plastic mint gum container or a small package of your choosing (and/or availability).

Step 9

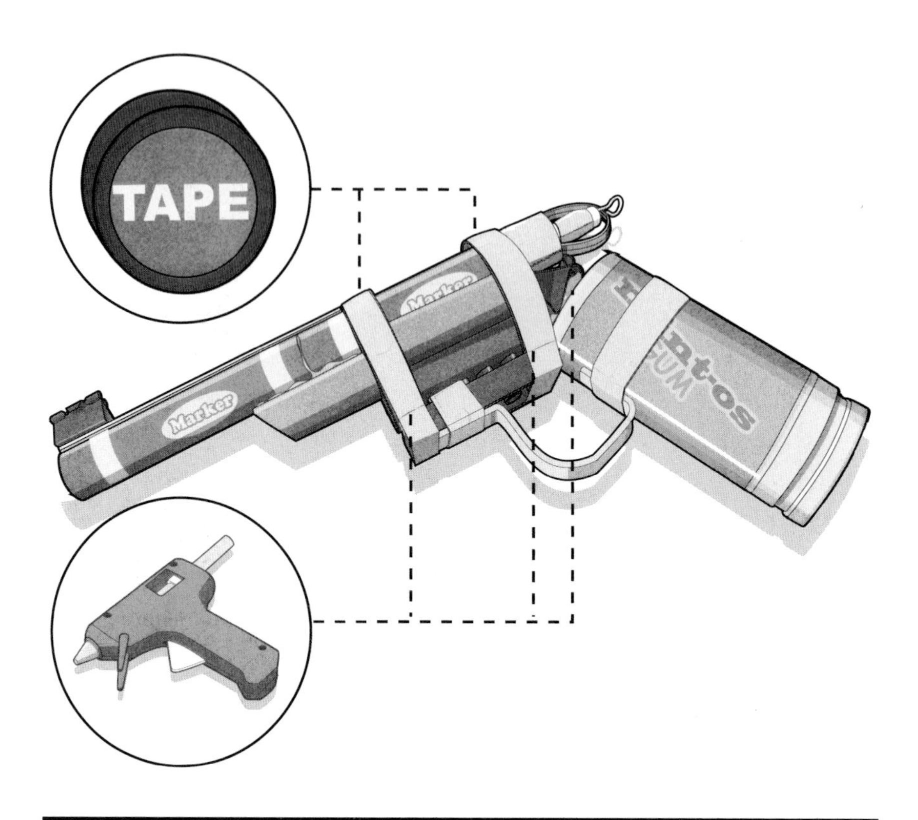

Now combine the grip assembly with the barrel assembly using hot glue. For additional support, add duct tape around the milk jug caps and the building block, as shown.

Load the barrel and make sure the ammo slides back to the pen housing trigger. Pull back the elastic trigger and watch it fire.

This is a great MiniWeapon that can be customized and tweaked to perfection. A simple holster can be constructed from a cardboard tissue roll. To do this, cut two small belt slits into the side of the tube, then slide the 44 Marker Magnum into the tube.

ROUND-NOSED "BULLET"

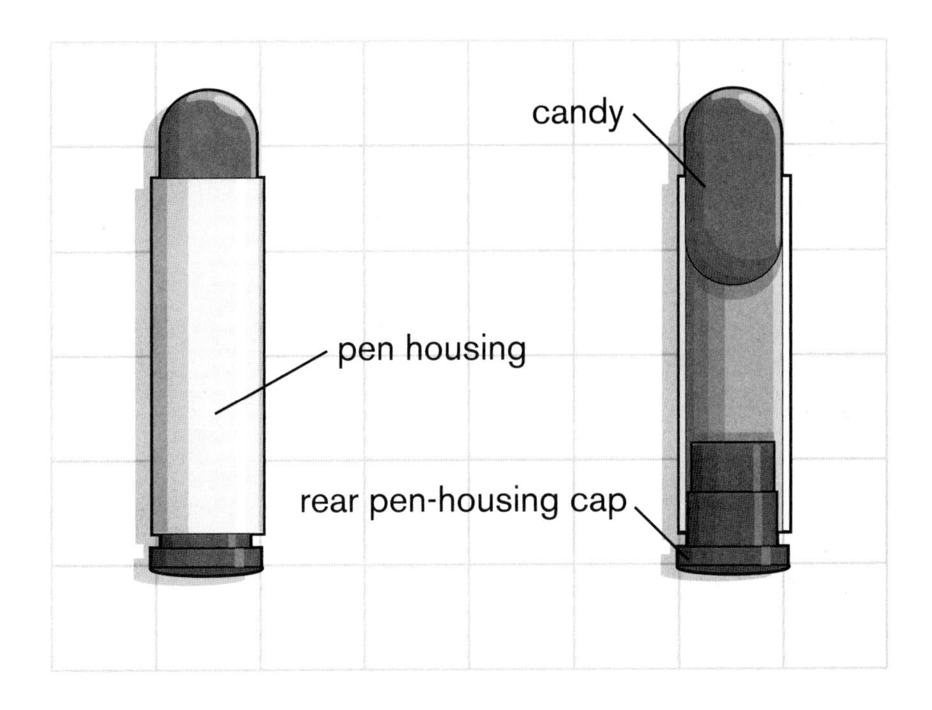

Most MiniWeapon launchers have several ammunition options, from small candies to mini marshmallows, but a Round-Nosed "Bullet" gives the homemade weapon enthusiast something authentic. What this ammo lacks in firing distance, because of its size, it more than makes up for in awesomeness, something above and beyond a candy slug.

Supplies

1 plastic ballpoint pen
1 Tic Tac candy

Tools

Hobby knife

Step 1

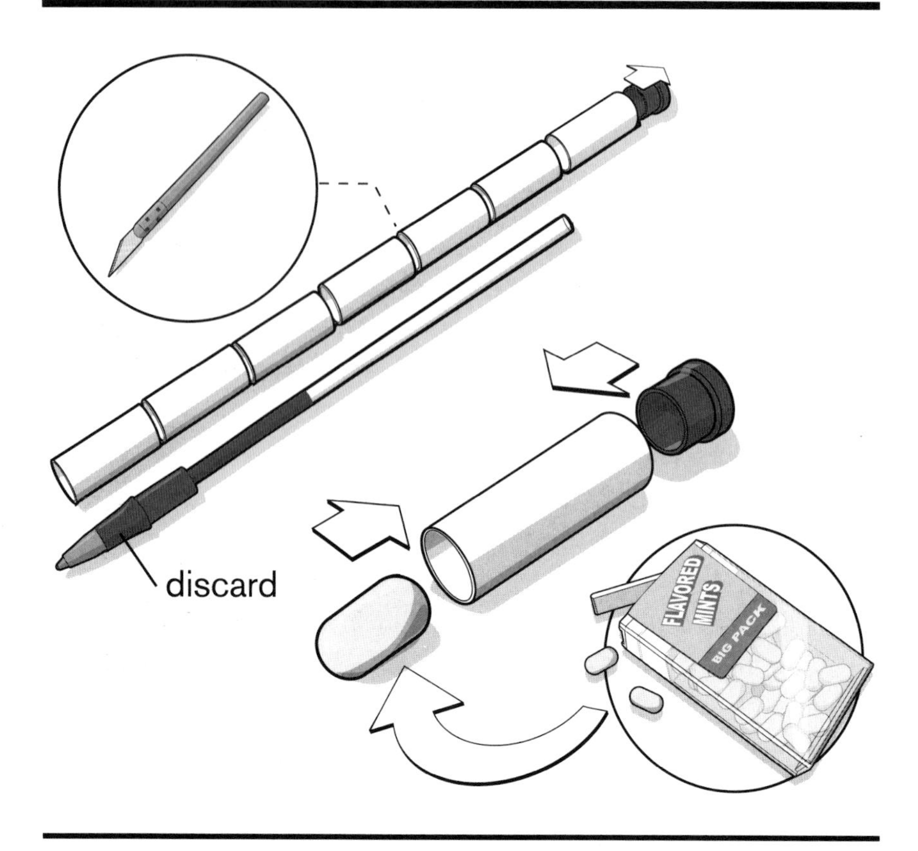

The Round-Nosed "Bullet" casing is manufactured from a simple plastic ballpoint pen. First, dissemble the pen: remove the rear pen-housing cap, using a hobby knife (or small set of pliers) if needed. Discard the ink cartridge; the rest of the pen's components will be reused.

Now, use a hobby knife to carefully slice off ¾-inch sections of the pen housing, as shown. Each section will form a bullet casing. Next, slide a Tic Tac into one end of a pen section. The fit will be tight, which is good. Push the candy only halfway into the pen housing.

To complete the bullet assembly, reinsert the rear pen-housing cap so it's not quite flush. This will be the rim. Rummage around your desk drawer to find a few more pen caps to complete a few more rounds with the leftover ¾-inch sections.

Never aim a loaded MiniWeapon launcher at a human or animal! However unlikely, the Round-Nosed "Bullet" can shatter on impact. Always expect the unexpected.

2

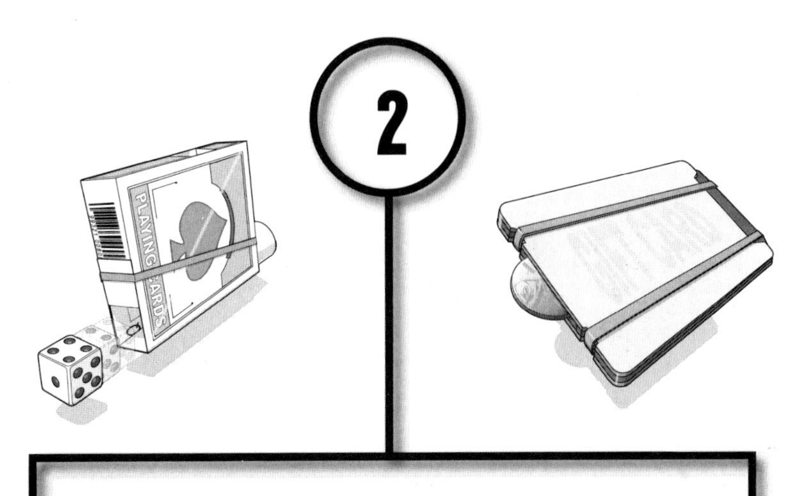

WEAPONS IN DISGUISE

SEMIAUTO DICE LAUNCHER

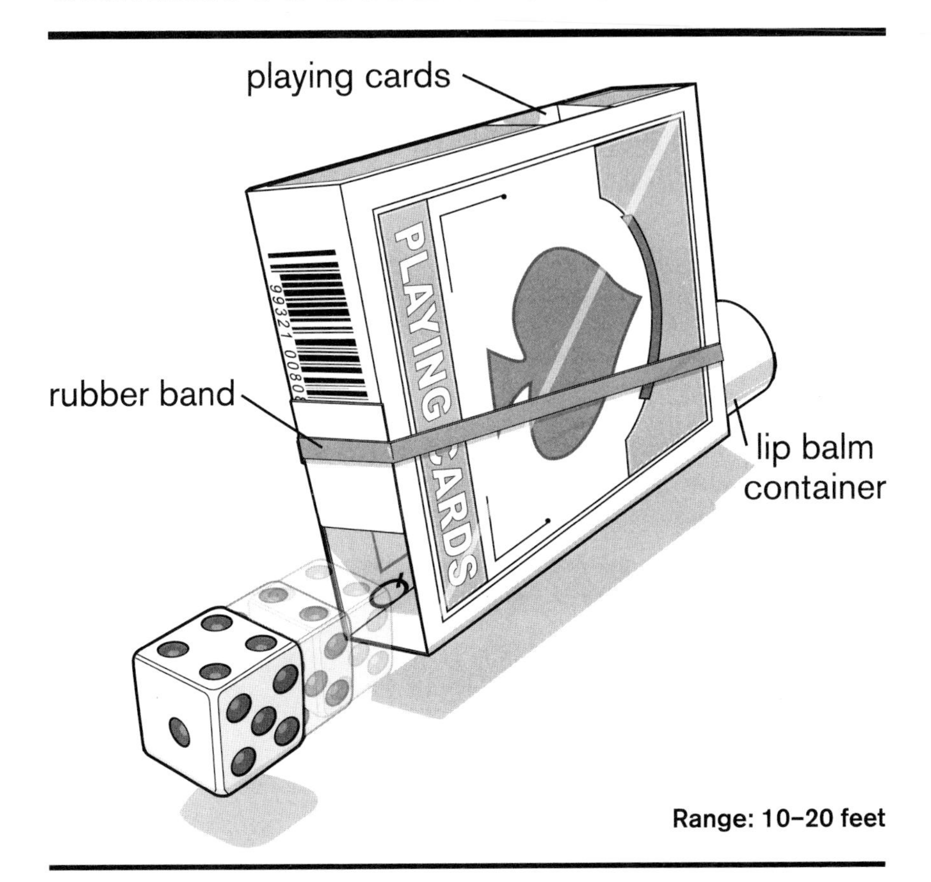

Range: 10–20 feet

Whether Double-O or Single-O, the life of a spy is high stakes. Secret missions often put you in close proximity of the intended target—code name: Rabbit. Gambling to keep your true identity shielded takes skill and training. The good news: the Semiauto Dice Launcher is the perfect MiniWeapon to catch your "royale" target off guard.

Supplies

1 deck of playing cards with box
1 lip balm container
1 wide rubber band

Tools

Safety glasses
Scissors
Superglue or hot glue gun
Hobby knife

Ammo

3+ dice

Step 1

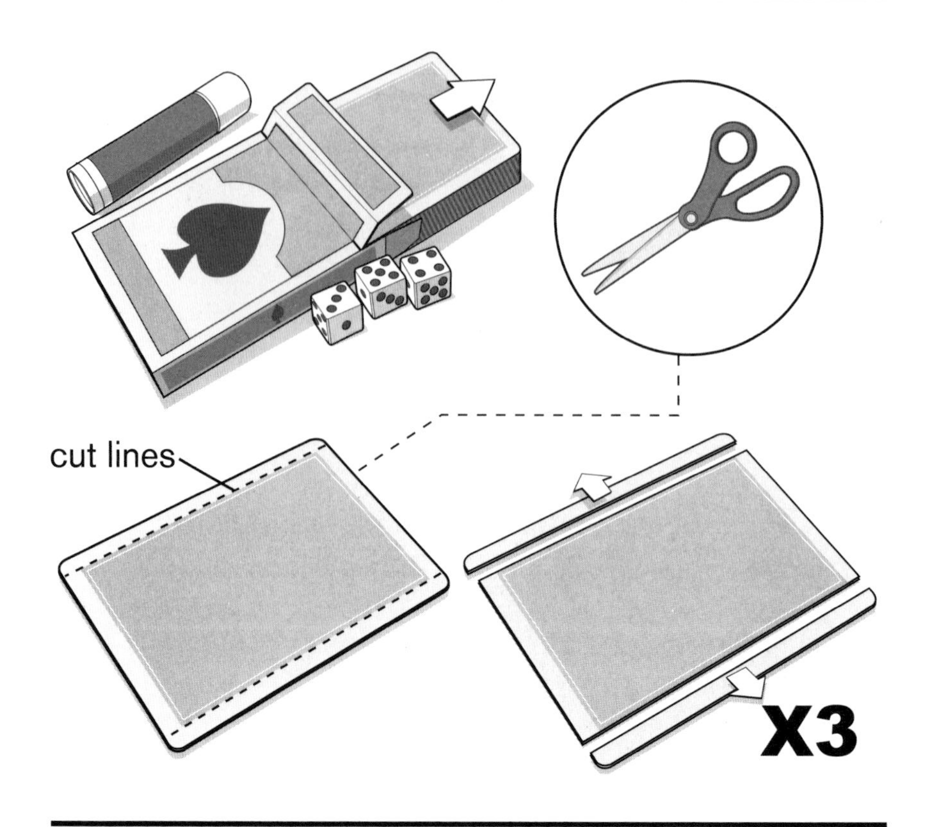

Gather your materials. Using scissors, trim off approximately 3/8 of an inch from the two long sides of a playing card, then repeat that step for two more cards. *Do not* cut the short sides. Depending on the design on the back of the cards, you might be able use the design border as a guide.

Step 2

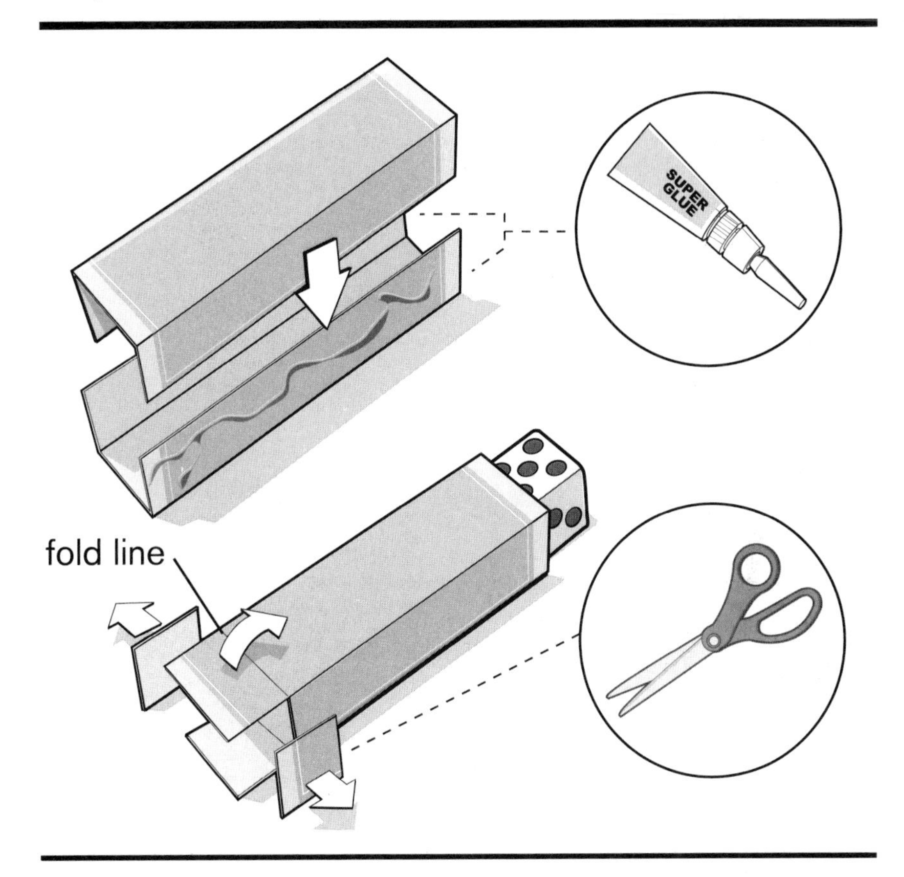

Now fold two of the cut playing cards twice to make two three-walled channels. The center distance between crease lines should be the width of the card box–use the box as a guide, then fold the cards. Dry fit the two cards together to create a tunnel. Test to see if the dice can slide through the tunnel easily, then superglue or hot glue them together.

On one end of the card tunnel, cut four ½-inch slits up each of the four corners. Then fold out the four sides at 90-degree angles (not shown). Cut off two of these flaps, opposite one another, so that only two of the flaps remain; refer to the illustration.

Step 3

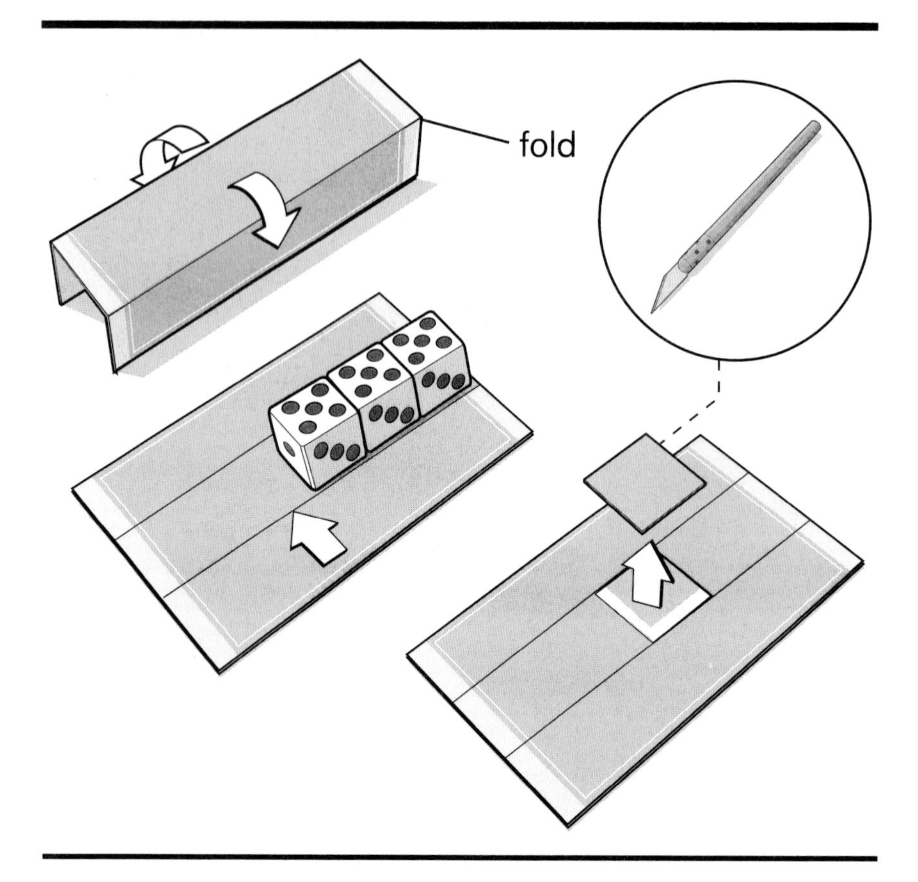

Fold the third cut card into a three-sided channel, similar to the last two cards. The center distance between crease lines should be the width of the playing card box.

After folding, unfold the card and lay it flat. Using the dice as a template, place three of them together down the center section of the playing card, starting from one end as shown. Mark the square "footprint" of the innermost die and remove that square with a hobby knife.

Step 4

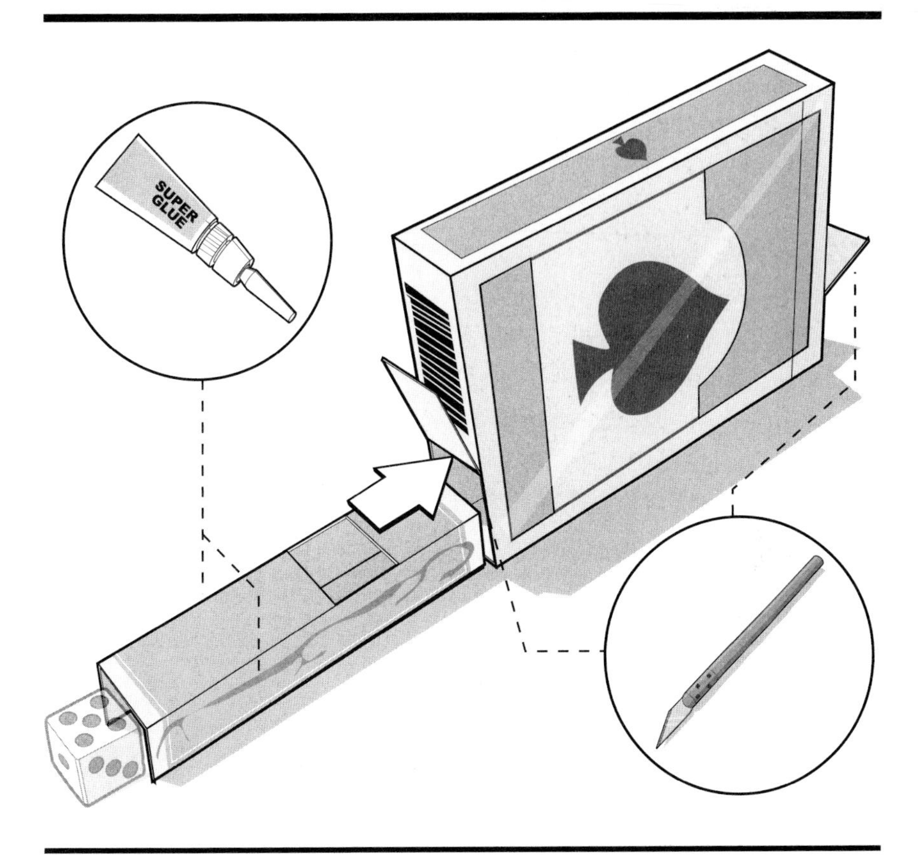

Next, cut two small doors on the bottom of the playing card box. Both door openings should be approximately the height of the last folded playing card.

Apply glue to the two sides of that card, then carefully slide it into the card box with the trough opening down, as shown. Insert your finger and press the card walls and the box walls together. The card walls should rest flush on the bottom of the box, and the dice must be able to slide through this square channel unobstructed.

Step 5

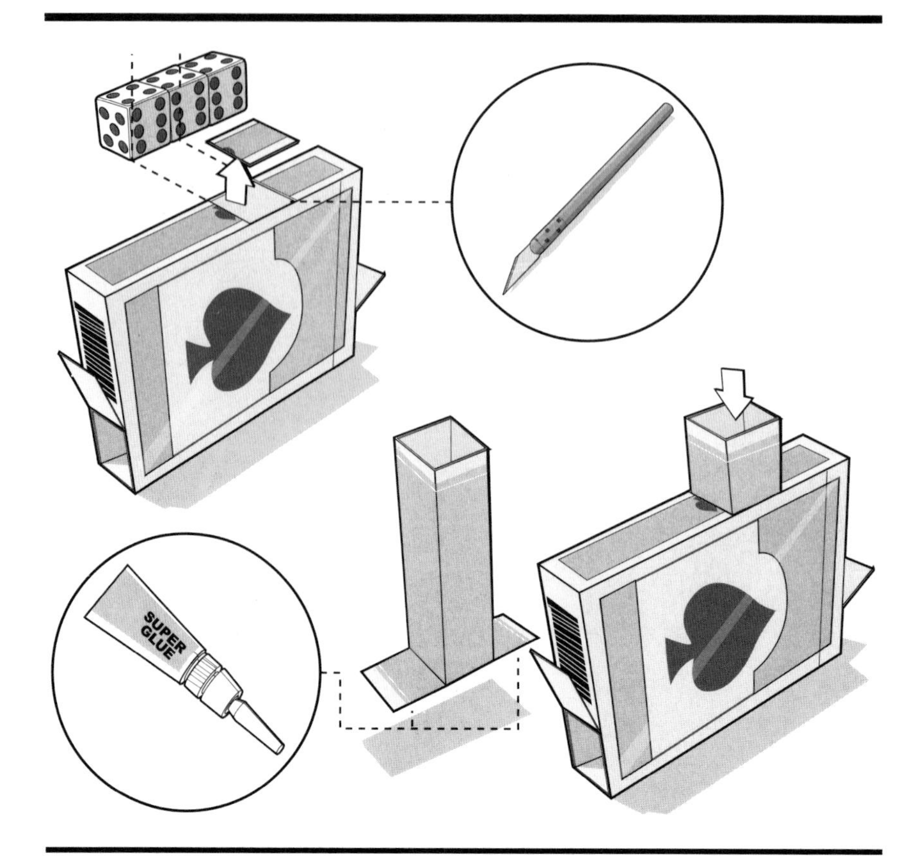

Now you need to create another hole through the top of the card box that aligns with the die-sized box you removed from the playing card channel earlier. Use three dice to measure the exact distance, and then cut out the box material with a hobby knife.

Apply glue to the two flaps on the custom assembly made out of playing cards, then slide the card assembly into the hole you just cut out. Connect the card assembly to the card slot already fastened inside the box. Once the glue is dry, drop a few dice into the slots to confirm they can pass through both corridors, down from the top and end to end.

Step 6

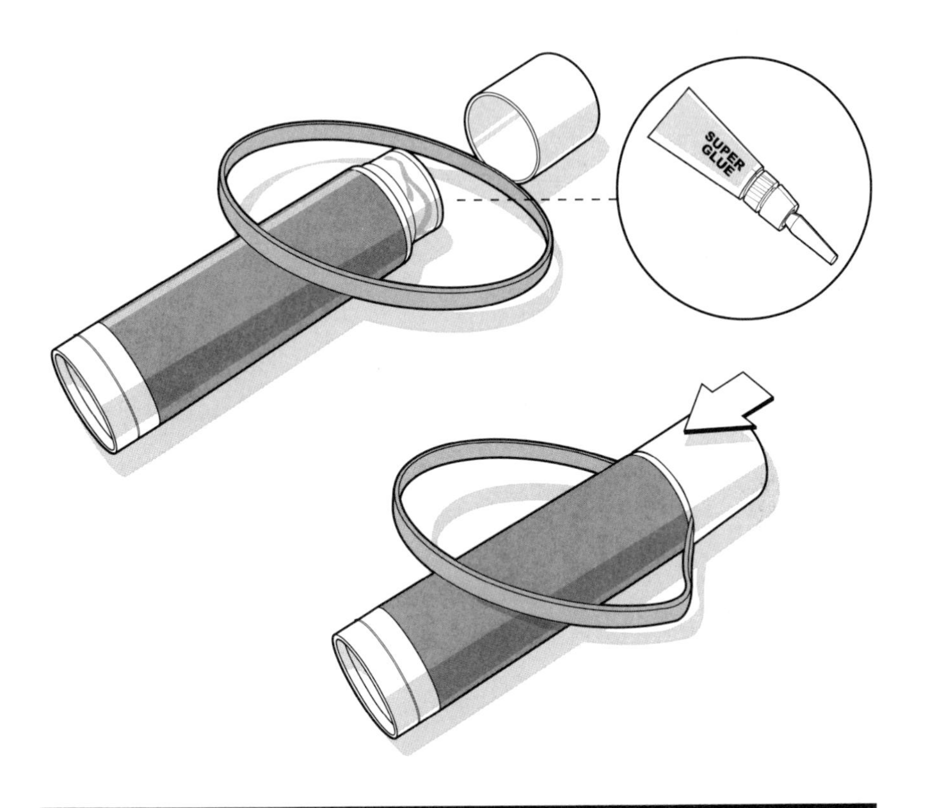

Next, remove the cap from a lip balm container and coat the inner, sealing surface of the container with a little superglue or hot glue.

Place a wide rubber band across the top opening and squeeze the cap back onto the lip balm container. The rubber band will be sandwiched between the cap and container, sealing it into place.

Step 7

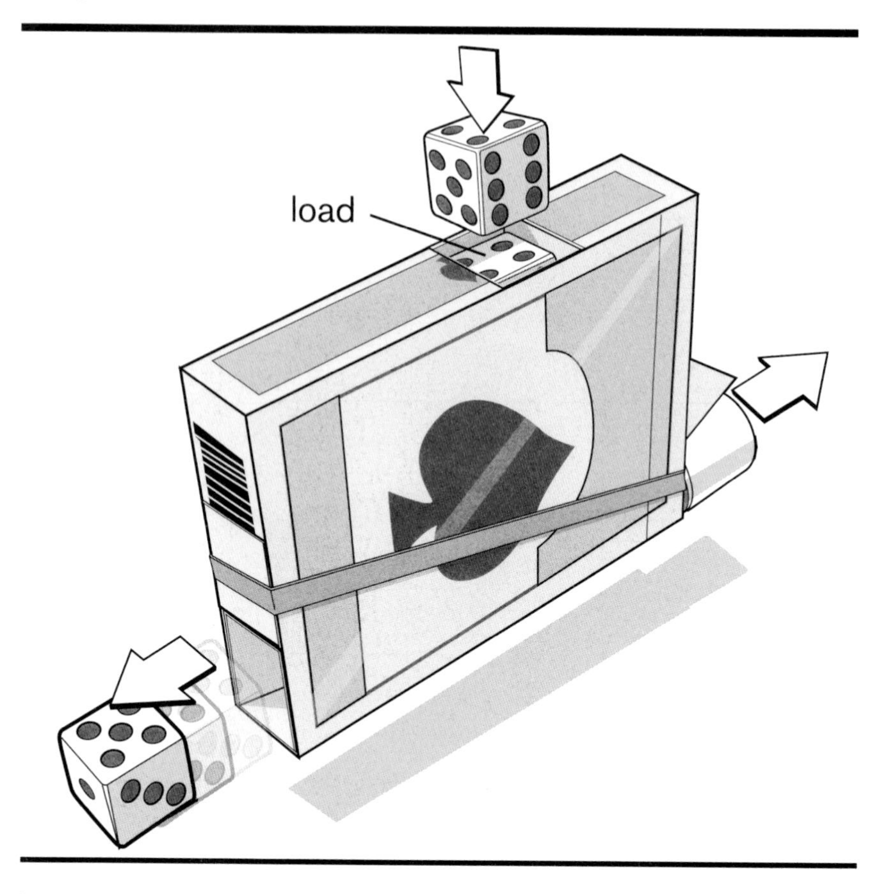

Insert the lip balm into the rear door of the playing card box, and wrap the rubber band around the box. You're finished!

Load several dice into the top hole, which is now your ammo clip. When you pull back the elastic lip balm trigger, one die will fall into the chamber. Release the lip balm container and let the die fly. Pull back the container once more to load and fire another die bullet.

To hide the weapon's identity, undo the rubber band and slide it into the box, then close the card box flaps. What could be sneakier?

GIFT CARD COIN LAUNCHER

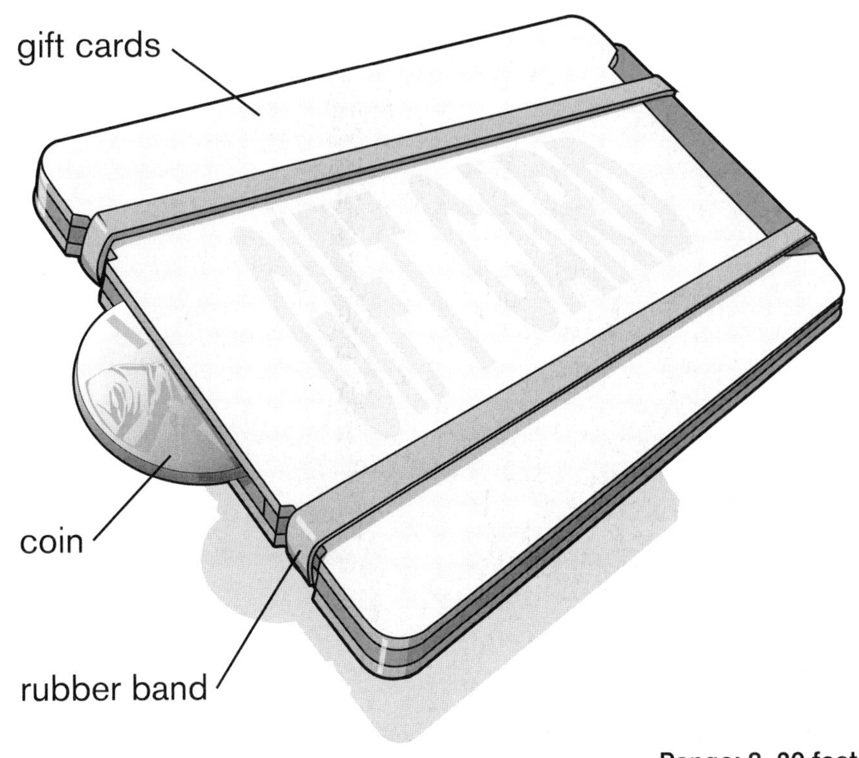

Range: 8–20 feet

Implementing the latest in MiniWeapon nanotechnology, the Gift Card Coin Launcher is paper thin, perfect for stashing in your pocket or wallet until it's time to make your target pay! Camouflaged with expired or zero-balance gift cards, this launcher will elude any security checkpoint. Just remember to fill your deep pockets with the kind of cash that jingles—not folds.

Supplies

3 expired or zero-balance plastic gift cards
2 heavy rubber bands

Tools

Safety glasses
Permanent marker
Hobby knife or scissors
Superglue or hot glue gun

Ammo

1+ coins

Step 1

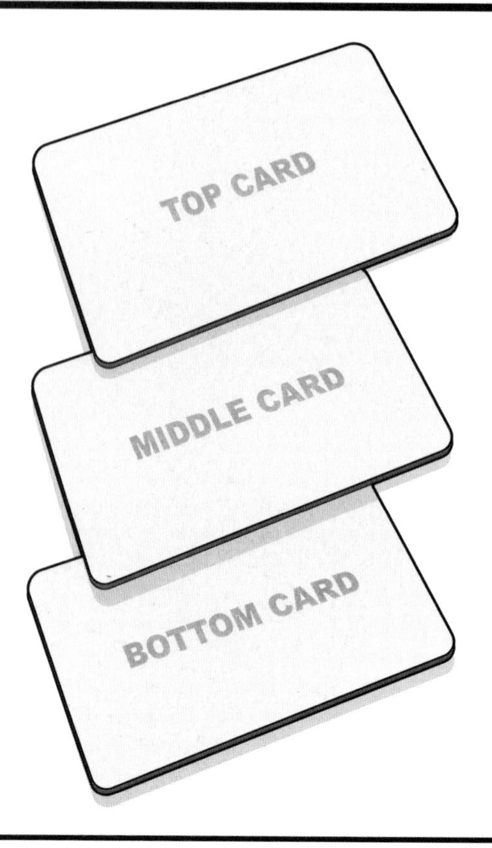

First, round up the gift cards. Rummage through the junk drawer or junk mail to locate three expired or zero-balance plastic gift cards. Promotional, nonfunctioning plastic credit cards or unused frequent flyer cards also work. *All three cards will be glued together or cut, so do not use a card that is still active.*

Step 2

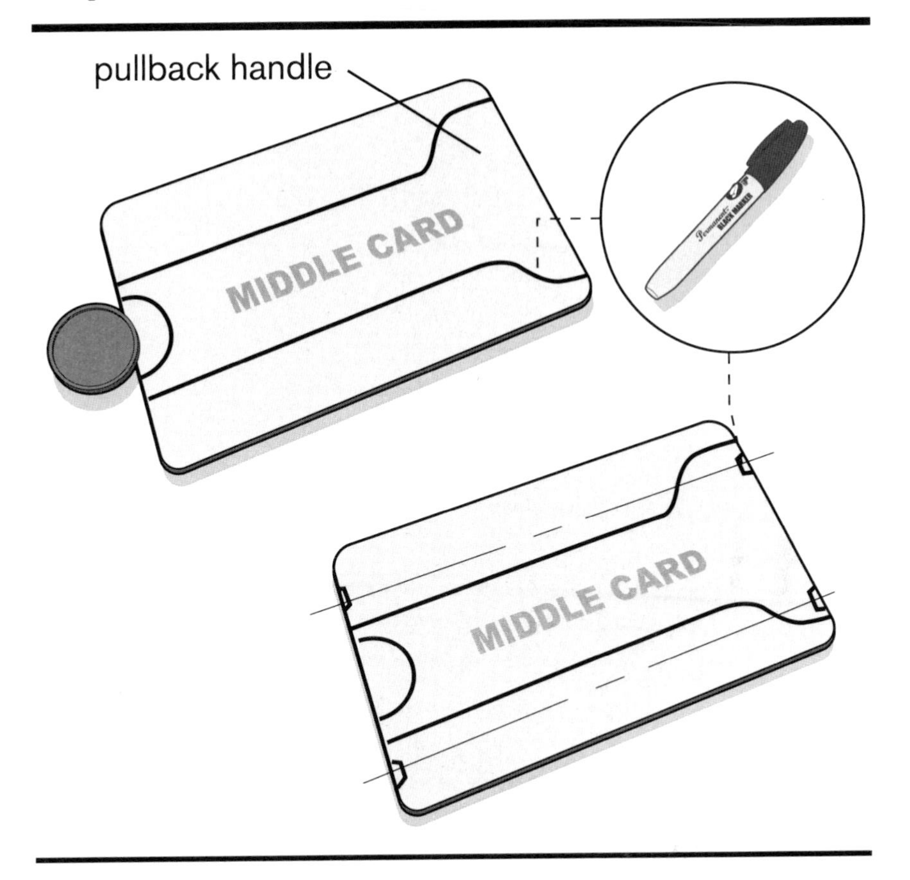

The middle card must be transformed into the firing mechanism—the most complicated part of this project.

Center the coin you will use as ammo halfway onto the middle card, then trace the diameter using a permanent marker. Use the illustration for reference; the diameter can vary depending on the coin ammo yet yield the same results.

On the same card, draw two lines parallel to one another slightly offset from the half-coin outline. As the two lines near the opposite end of the coin template, increase the center width. This wider end will eventually be the pullback handle.

Cut four small slots approximately the same width as the two heavy rubber bands. These four slots should be parallel. Notice on the illustration how two of the marker slots are on the firing mechanism (center piece) while the other two are outside the parallel lines (outer pieces); this will help you determine how wide to make the pullback handle.

Step 3

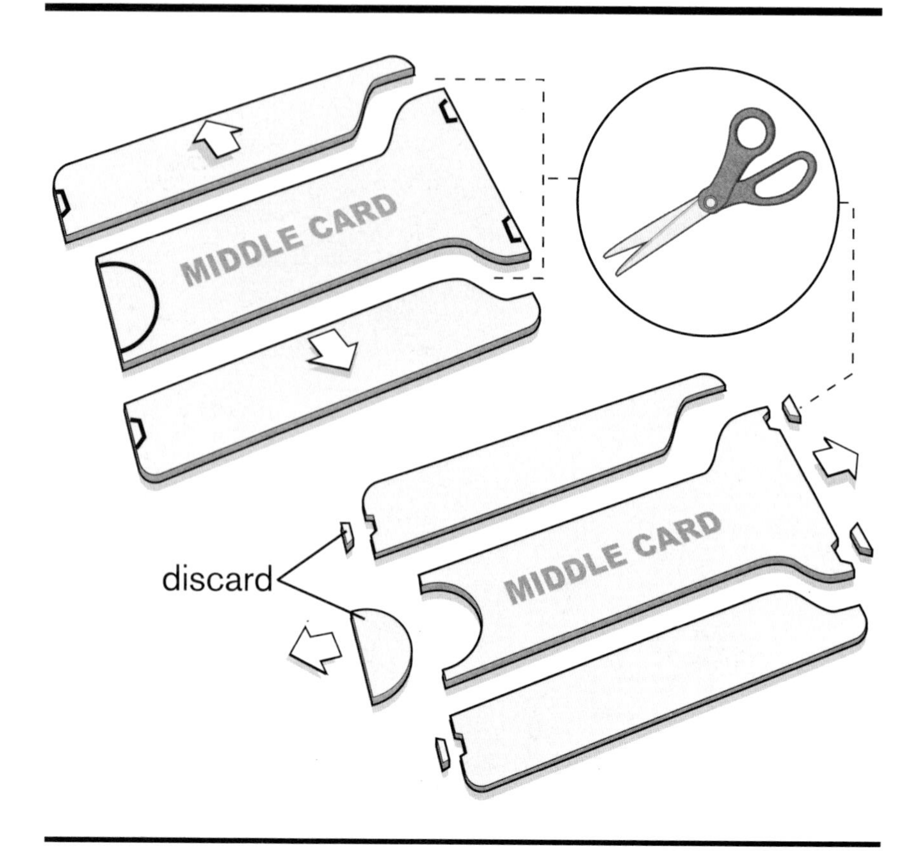

Once your marks match the drawing, use scissors to separate the card into three sections by following the two parallel marker lines. Don't forget to "S" cut out the pullback handle.

Next, remove the semicircular coin piece from the center section. Discard the half circle. Then remove the four very small tabs. These tabs are approximately the width of the rubber bands. Discard the tab scraps.

Step 4

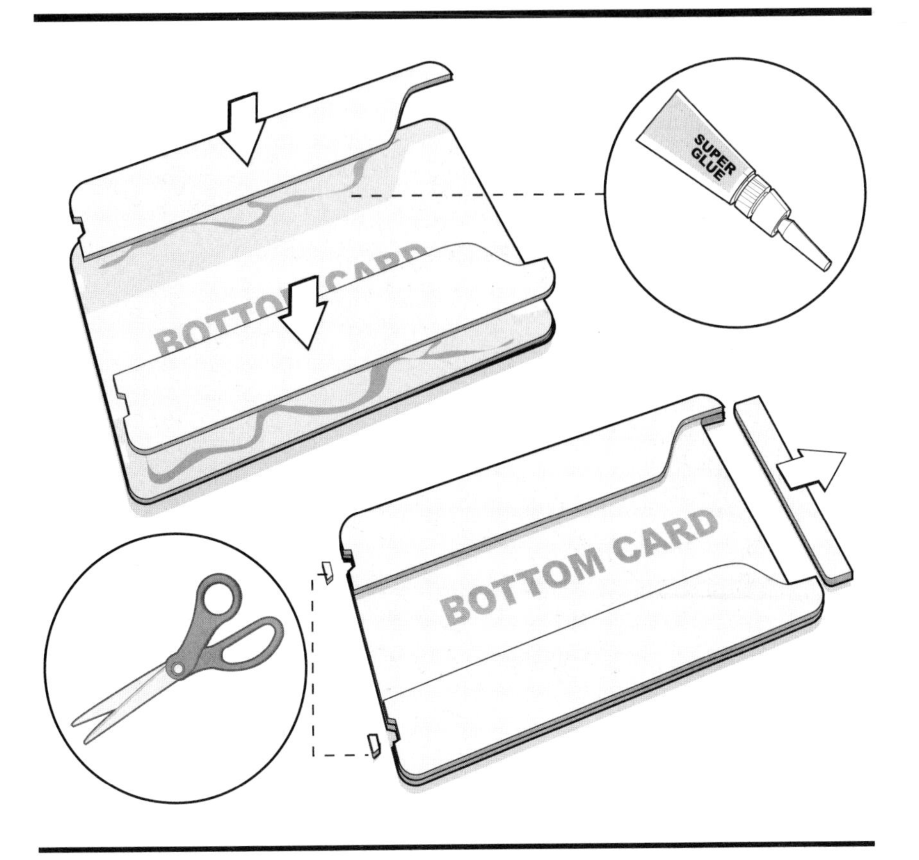

Using superglue or a hot glue gun, fasten the two outer pieces of the middle card you just crafted onto the top of the bottom card. The sides should be flush to the edges. Be careful not to spill any glue onto the center of the card–any glue in the center will hinder your launcher's performance.

After the glue cools or dries, remove the same two tab notches on the bottom card using the newly placed card sides as a template.

By the pullback end, remove ¼ inch of plastic card from the bottom card. Use the illustration as a reference. Removal of this material will allow you easily to access the pullback trigger with your fingers.

Step 5

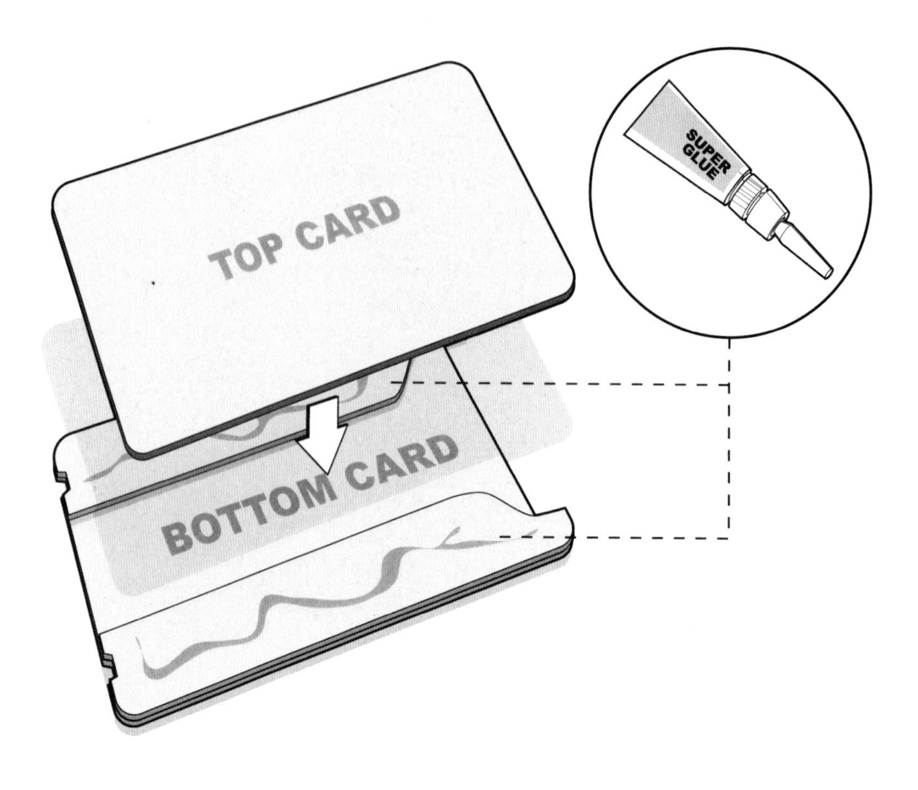

Now carefully glue the top card onto the bottom-and-middle-card assembly. Remember not to overglue—nothing should ooze into the inner channel. Before the glue sets, make sure all the sides are flush.

Step 6

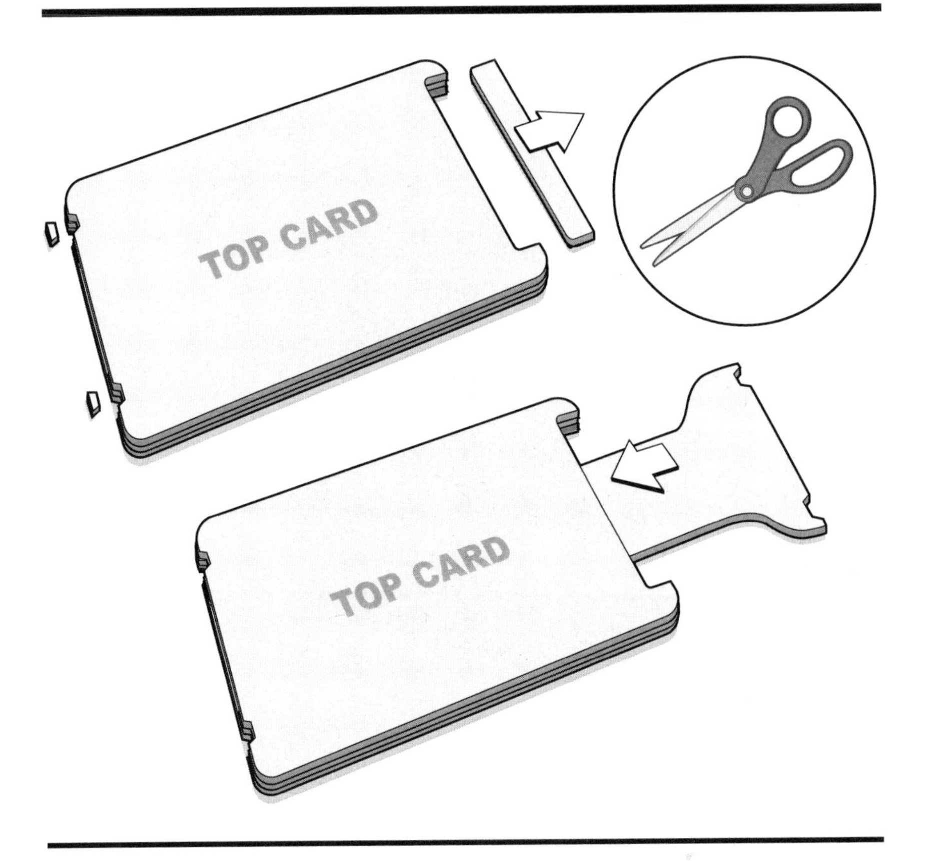

At this point, all three cards should be sandwiched together with glue. Wait till they have completely dried before continuing.

Again, you must remove another small set of rubber band tabs from the third gift card using scissors or a hobby knife. Use the other two card slots as a template.

Also, remove approximately ¼ inch of material by the pullback area. Use the bottom card cut line as a reference.

Slide the final middle piece into the middle channel. If it rubs the sides and doesn't go all the way in, pull it back out and shave a small amount off the outer edge to make it fit.

Step 7

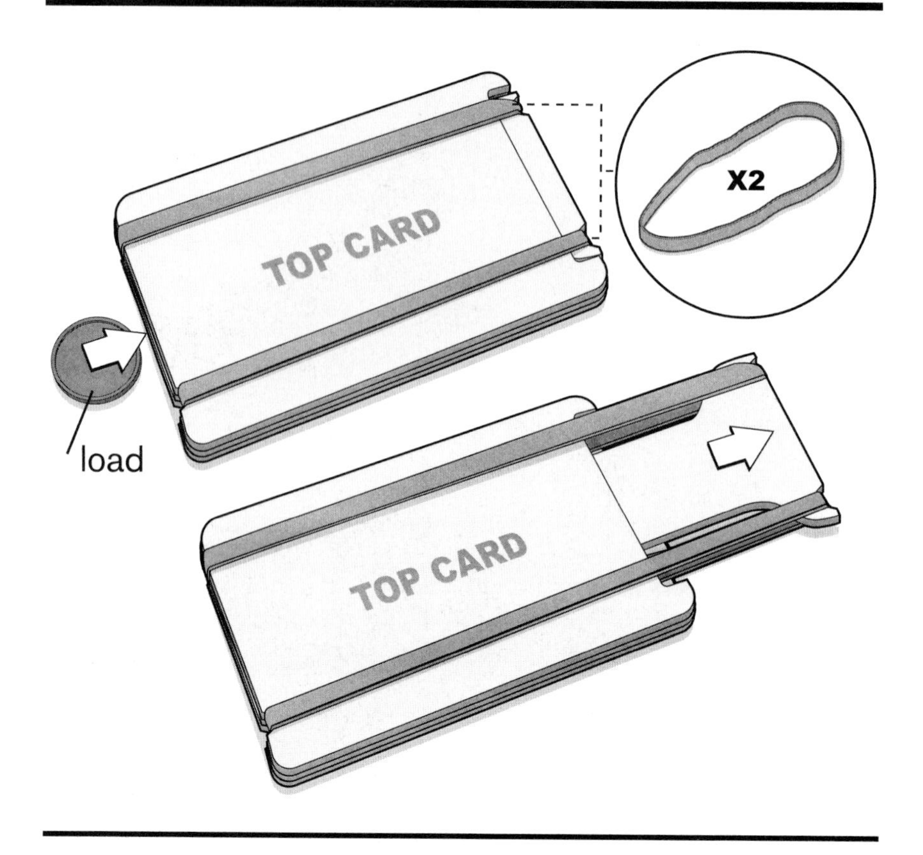

Now it's time to attach the power supply! Slip two heavy rubber bands over the full assembly and slide them into the four small end tabs you cut out earlier. The bands should be tight and incapable of falling off. If this is not the case, substitute both of them with a smaller size.

Now pull back the plunger and load the coin ammo into the front slot until the coin is no longer visible. Then carefully pick your target and let it rip! The coin will be catapulted out the front of the launcher and travel at a high velocity until it kisses its target.

Firing coins can be dangerous and damaging. Always use common sense when picking a nonliving target, and use this MiniWeapon at your own risk.

MINT TIN CATAPULT

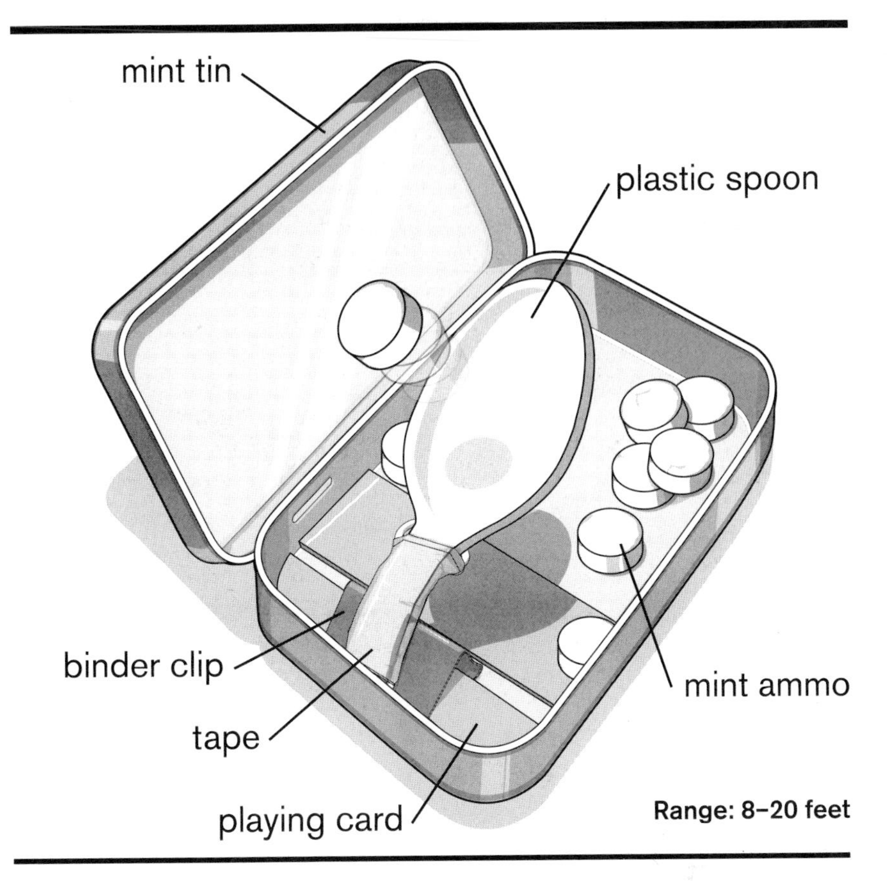

Tucked away in this small tin is a powerful weapon capable of launching breath mints almost 20 feet. The Mint Tin Catapult has been redesigned with a single binder clip launching arm, making its construction and durability simple and impressive. How you wield such a MiniWeapon is your choice, but you'll deliver your foul-mouthed targets a mint they won't soon forget!

Supplies

1 plastic spoon
1 mint tin (Altoids or similar)
Duct tape
1 medium binder clip (32 mm)
1 playing card

Tools

Safety glasses
Hobby knife or scissors
Hot glue gun

Ammo

1+ soft candies

Step 1

First, use scissors to cut the handle of a plastic spoon so that the spoon can be concealed inside the tin. Use the tin's interior length to find the approximate length, then trim the spoon with scissors. Discard the handle scrap.

Next, securely duct tape the short-handled spoon onto the metal handle of a medium binder clip. Slightly flex the spoon to confirm that it's secure.

Step 2

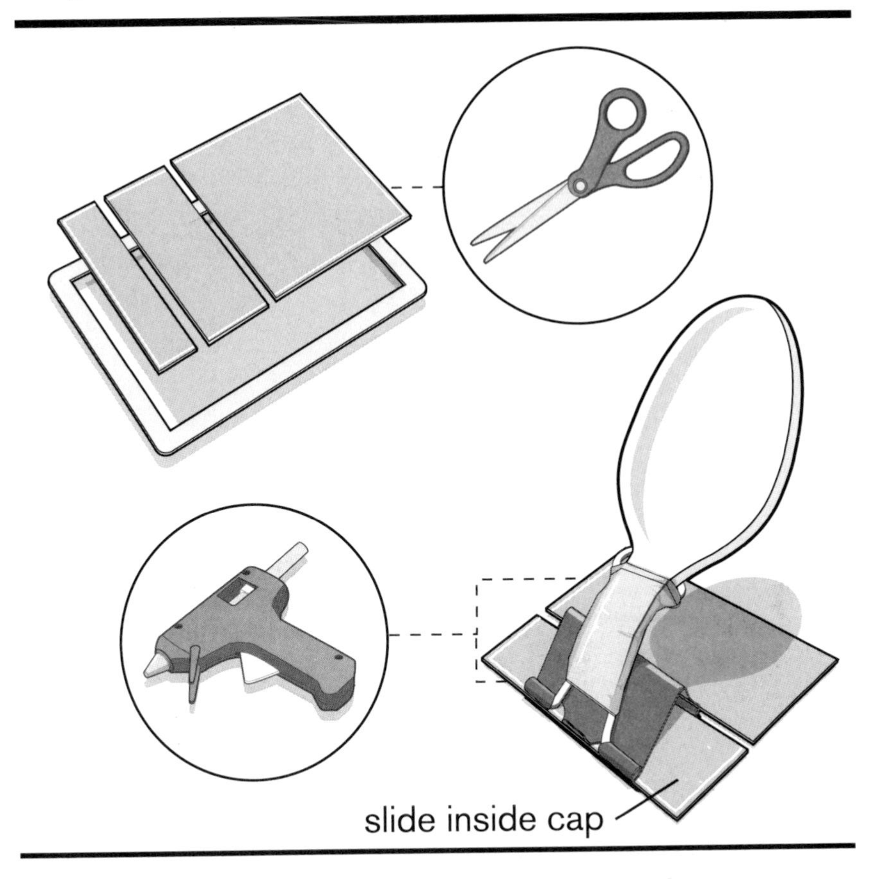

To support the power of the binder clip throwing arm, you'll use two playing card segments to help anchor it to the tin container's floor. If a playing card is not available, use a material similar in thickness and durability.

Cut the playing card to the same width as the interior width of the mint tin. Discard the extra clippings.

Use scissors to cut off the first section of card; this section should be the same width as the binder clip. Once it's cut, glue the card segment to the *interior* of the clip as shown in the illustration.

The second card segment should be longer than the metal binder clip handles. Once cut, glue the card onto the *top* surface of the handle, as shown.

Step 3

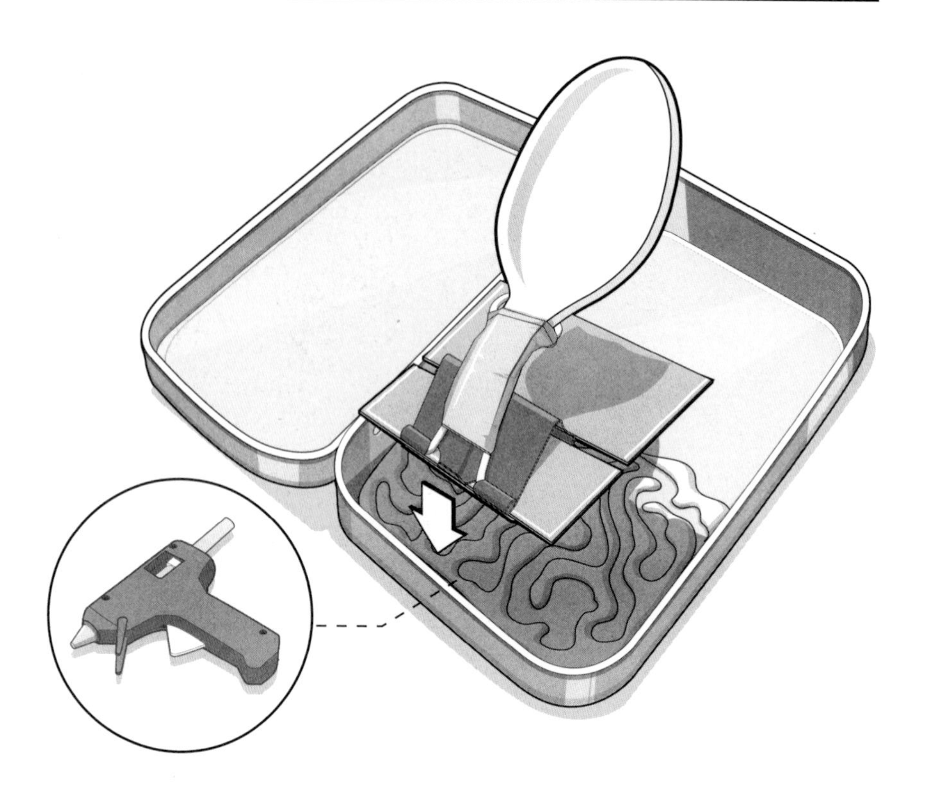

Now that your catapult's launcher assembly is complete, it needs to be fastened to the interior of the tin container. Load the interior floor with hot glue, and before it can dry, place the catapult assembly on top and press it into place. Once the glue cools, you're ready to launch.

The tin wall will help stop the catapult arm from firing mints into the ground. Adjust or modify it for desired results. The tin is also a good place to store your ammo.

Remember to use eye protection! Never aim this catapult at a living target and use only safe ammunition. Soft mints and mini marshmallows work nicely.

CRAYON CANNON

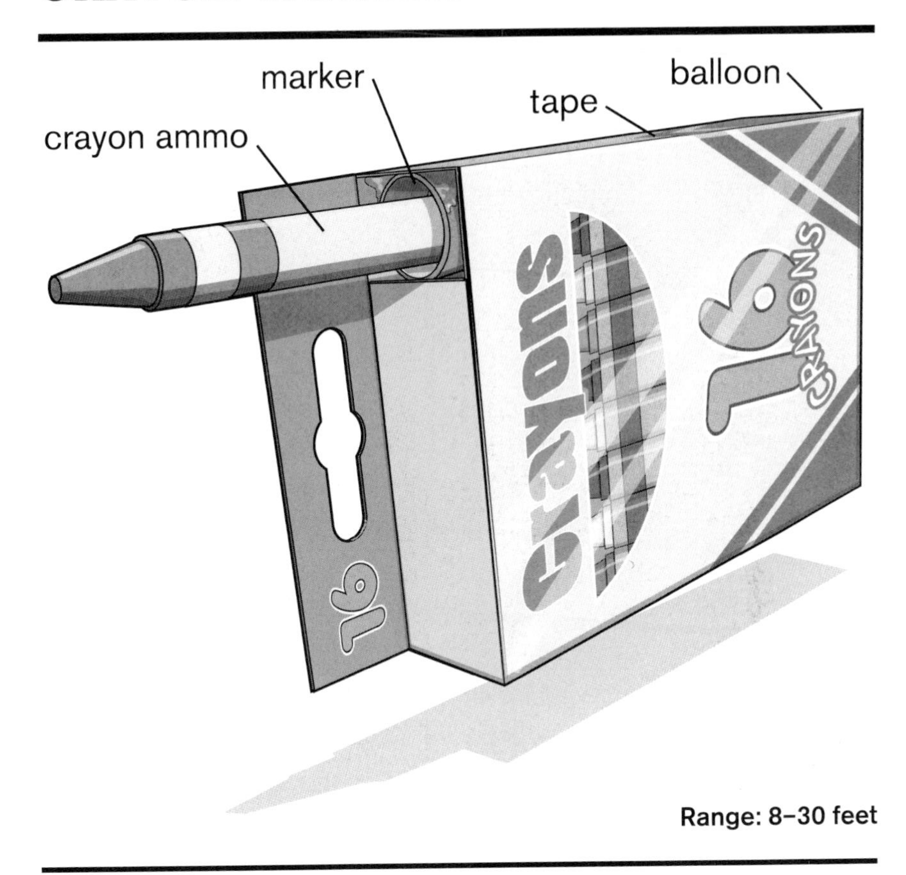

Range: 8–30 feet

Let's hear it for the Crayon Cannon! Loaded with 16 colorful rounds of ammunition, this single-shot launcher is the perfect MiniWeapon for the artistic sharpshooter. Boxed in an all-in-one artillery housing, it can be transported during undercover operation and is a cinch to refill. Before unloading, pick your target carefully, or the writing will be on the wall!

Supplies

1 crayon box (16 count)
1 standard art marker
1 small balloon
Duct tape

Tools

Safety glasses
Hobby knife
Pliers
Hot glue gun

Ammo

1+ crayons

Step 1

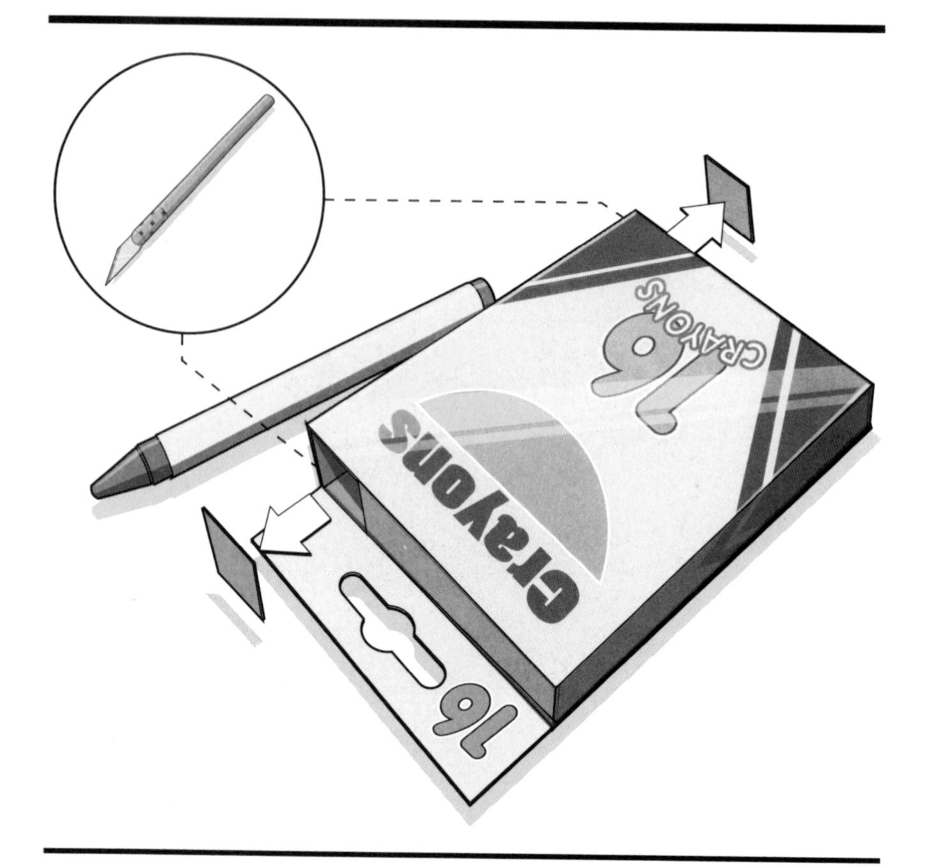

In this version of the Crayon Cannon, you'll be using a 16-count crayon box because of the box's width. Other boxes can be adapted from these general instructions if this size of box is not available.

Empty—but do not discard—all 16 crayons from the box. Then use a hobby knife to remove two cut squares from the top cover and bottom of the box. The dimensions of these squares should be slightly larger than those of a standard art marker.

Step 2

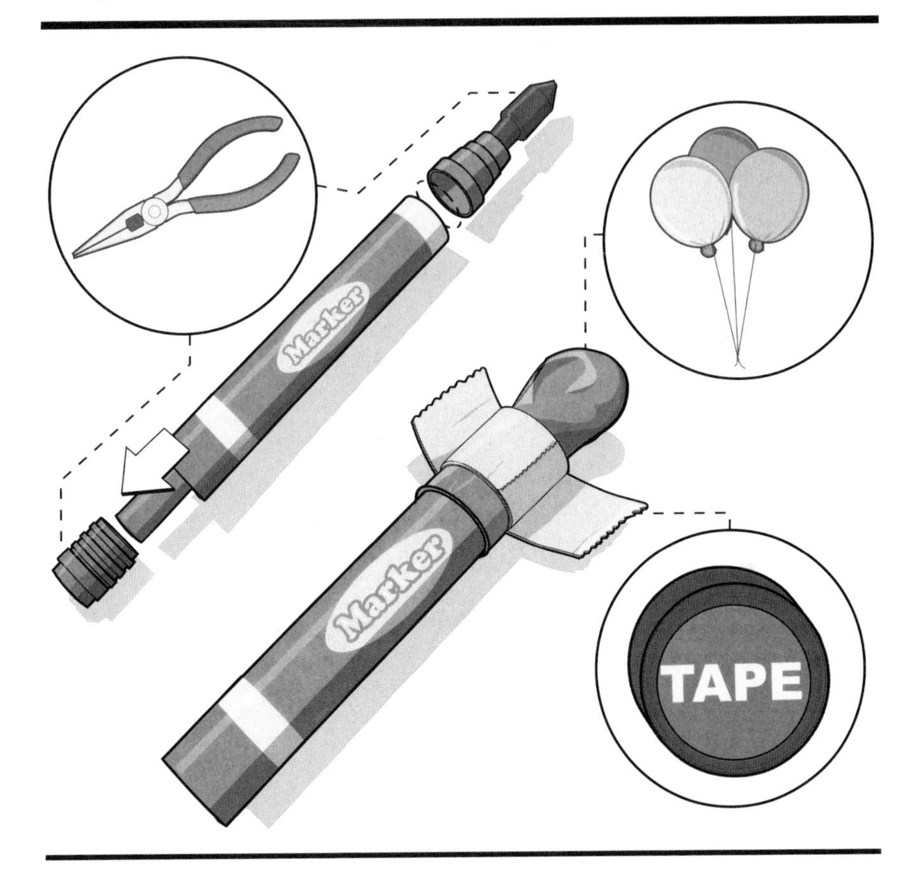

The barrel of the Crayon Cannon is fabricated from one art marker. If possible, a recycled plastic marker is preferred, but any marker will work. Using pliers, remove the marker's end cap by slightly rotating it from the housing. Again, use the pliers to remove the marker nib by sliding it out the tip of the marker. Discard this nib to avoid a mess.

Using a hobby knife, pierce the housing below the plastic point on the marker housing and slowly rotate the marker. This will make cutting it safer. Once this end has been completely removed, slide out the internal ink cartridge and discard it. The housing pathway should be unobstructed by any plastic fragments that might have broken off during cutting. (If the housing contains ink residue, rinse and dry it.)

Now pull a small balloon over one end of the marker housing, with the majority of the balloon covering the housing and roughly ½ to ¾ inch overhanging the end. Tape the balloon securely into position, test the tape restraint a few times, and add more tape if needed.

Step 3

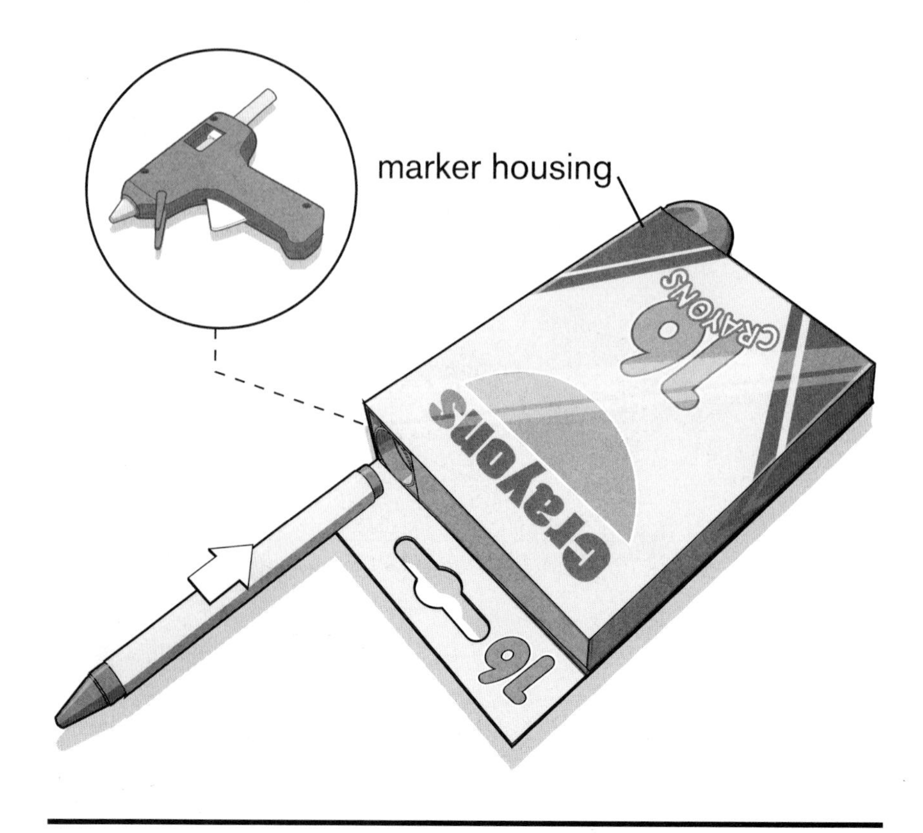

The last step is to install the marker barrel inside the crayon box.

Insert hot glue into both openings in the box you cut out earlier. Before the glue cools, carefully slide the marker into the box. Be careful not to melt the balloon with the hot glue.

Now load a crayon missile into the marker opening and let it drop down into the balloon. With your fingers, locate the bottom of the crayon in the balloon. Holding the crayon between your fingers, pull the balloon back and safely aim the cannon away from spectators and anything breakable. When ready, release the colorful ammunition from your grip and let it fly! The latex balloon will catapult the crayon missile through the barrel at high velocity.

Note: never operate the Crayon Cannon launcher if the balloon is showing signs of wear.

PEN BLOWGUN

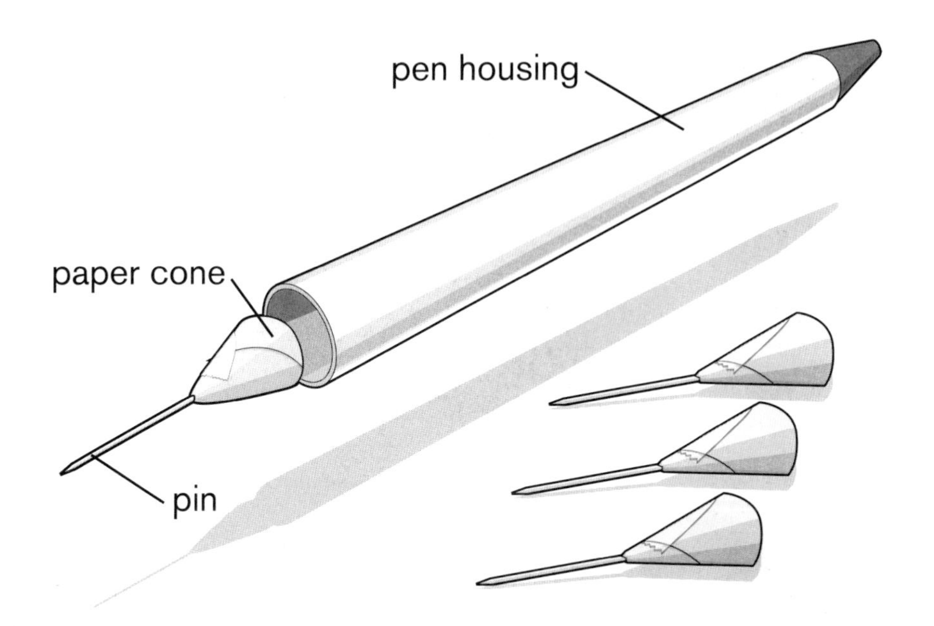

Range: 8–20 feet

Balloon targets beware! With lethal accuracy, and various dart designs, the Pen Blowgun will soon be a staple in your MiniWeapons arsenal. Designed with minimal construction, assembly is quick, perfect for an operative in training. Its pen housing makes the blowgun stealthy, untraceable by nonspies. The Pen Blowgun is capable of firing pointed and nonpointed darts, depending on the seriousness of your mission.

Supplies

1 plastic ballpoint pen
Clear tape

Tools

Safety glasses
Hobby knife or pliers
Scissors
Hot glue gun (optional)

Ammo

1+ sticky notes
1+ small metal pins or staples (optional)

Step 1

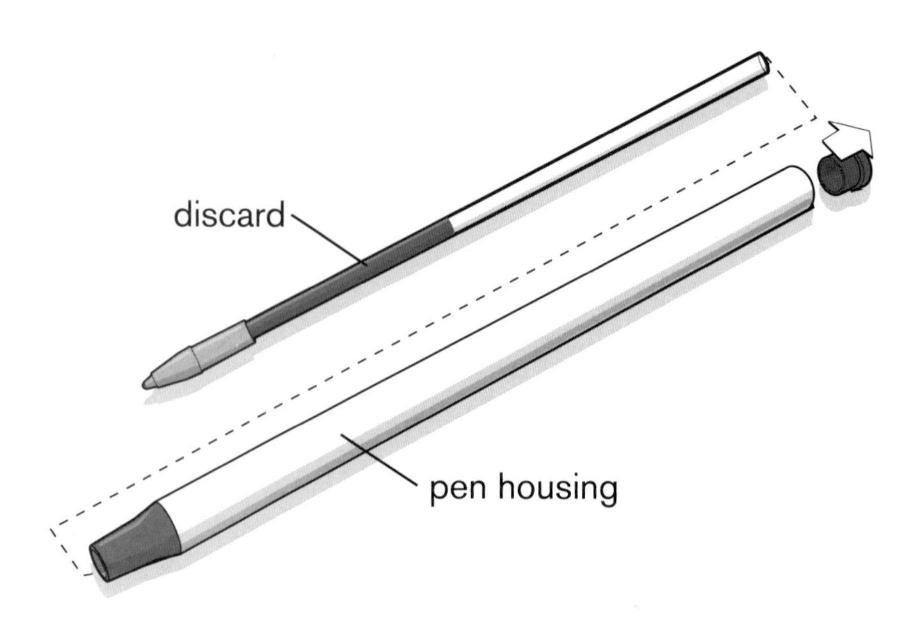

Disassemble one plastic ballpoint pen into its various parts. Depending on how the pen has been manufactured, you may need a tool to assist you with dislodging the rear pen-housing cap. You can use a hobby knife to cut it off or a small set of pliers to pull it out—both work just fine.

Step 2

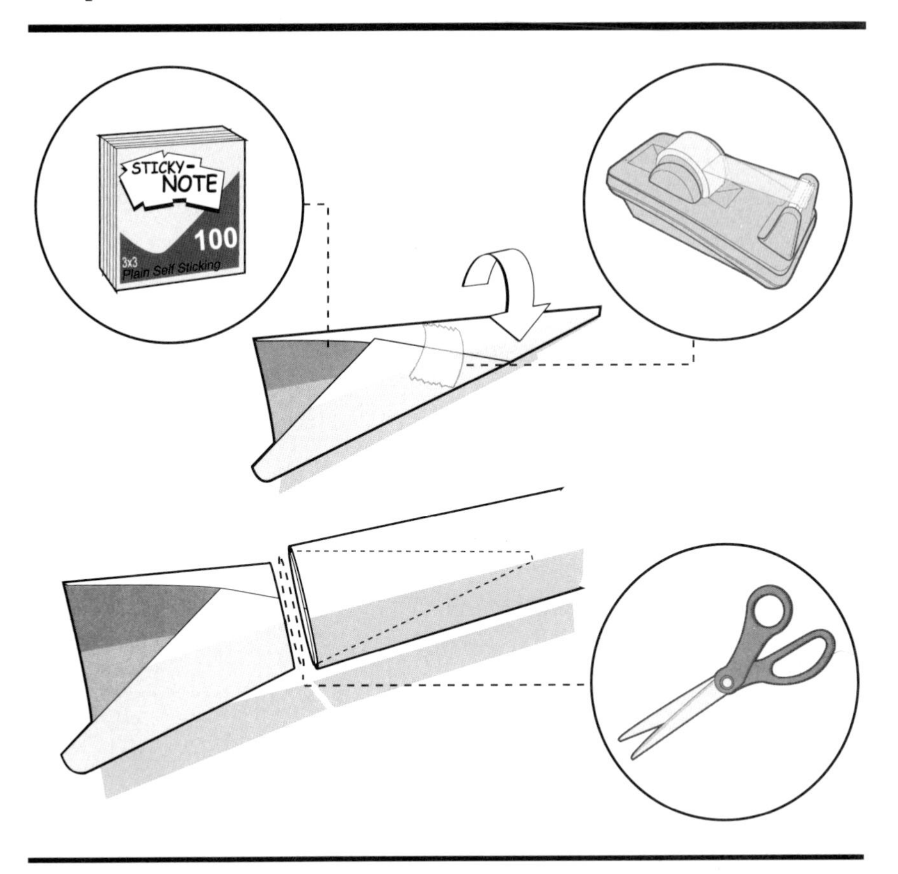

The blow darts are constructed from paper. Start with 1½-inch-square pieces of paper. If you have a 3-inch-by-3-inch sticky note available, fold it in half, then fold it one more time to create four stacked squares (not shown). Then unfold the sticky note and use the crease lines as your 1½-inch box cut lines (remove all four squares for four darts).

Now roll one of the small 1½-inch paper squares into a cone. Once you have a cone shape, place a small piece of clear tape on the edge to hold it in place.

Fit the small paper cone into the pen housing, being careful not to damage it. Once it's gently wedged into the pen housing, use scissors to trim off the extra material hanging out of the cylinder. Now the dart's maximum width is the same size as the pen tube's inner diameter. Repeat this cone construction for the other paper squares.

Alternate Construction

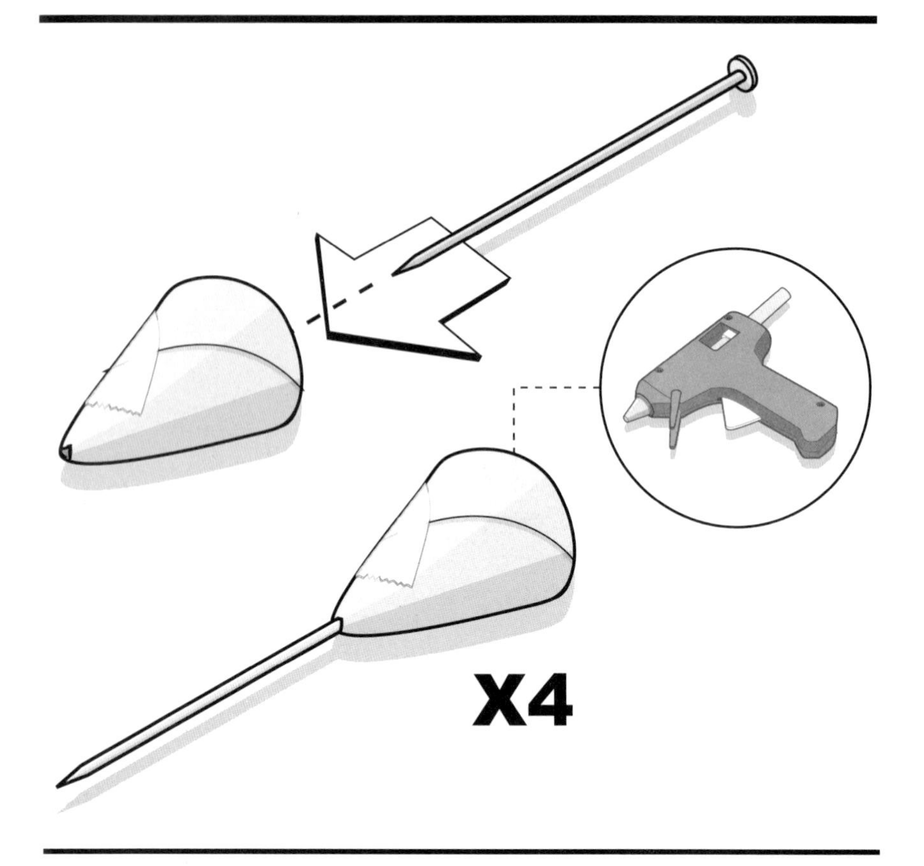

Your Pen Blowgun is complete! However, you do have the option to equip your dart with a straight pin or straightened staple if you're hunting balloons or dartboards.

Push the pin through the back of the cone so that the point protrudes from the pointy end. Once in place, dab a small amount of hot glue inside the cone to hold the pin in place.

Now stuff the dart into the pen housing with the point facing toward the exit. Take a deep breath, then aim the pen and hold it to your lips. Exhale sharply to blow the cone dart out.

It is very important to remember that you're firing needles from your mouth. ***You should never inhale while the pen is in your mouth waiting to be launched.*** Always be responsible and fire it in a controlled manor. ***Safety glasses are a must, as is staying clear of spectators.***

MINT PACK AND ARROW

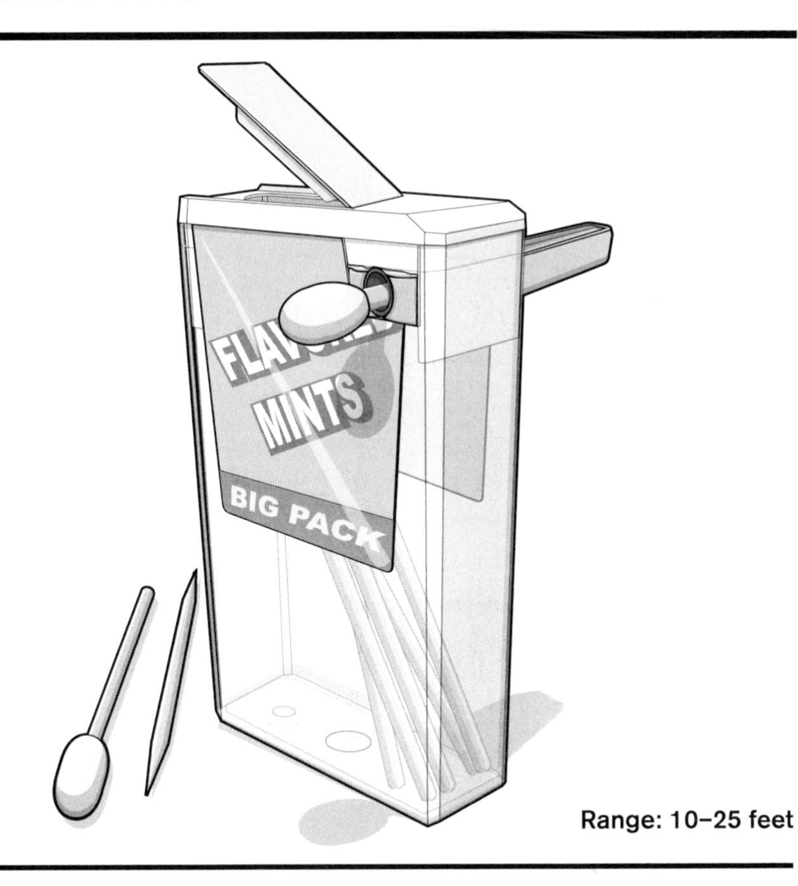

Range: 10–25 feet

Can I offer you a breath mint, General Evil? Super small and super cool, this all-in-one MiniWeapon launcher *pretends* to be a simple breath mint container but is capable of launching cotton-swab arrows 20 feet! Pocket-sized and fully stocked with ammo, the Mint Pack and Arrow packs a punch.

Supplies

1 inexpensive mechanical pencil
1 Tic Tac container
1 wide rubber band
Duct tape

Tools

Safety glasses
Scissors or hobby knife
Hot glue gun

Ammo

1+ cotton swabs

Step 1

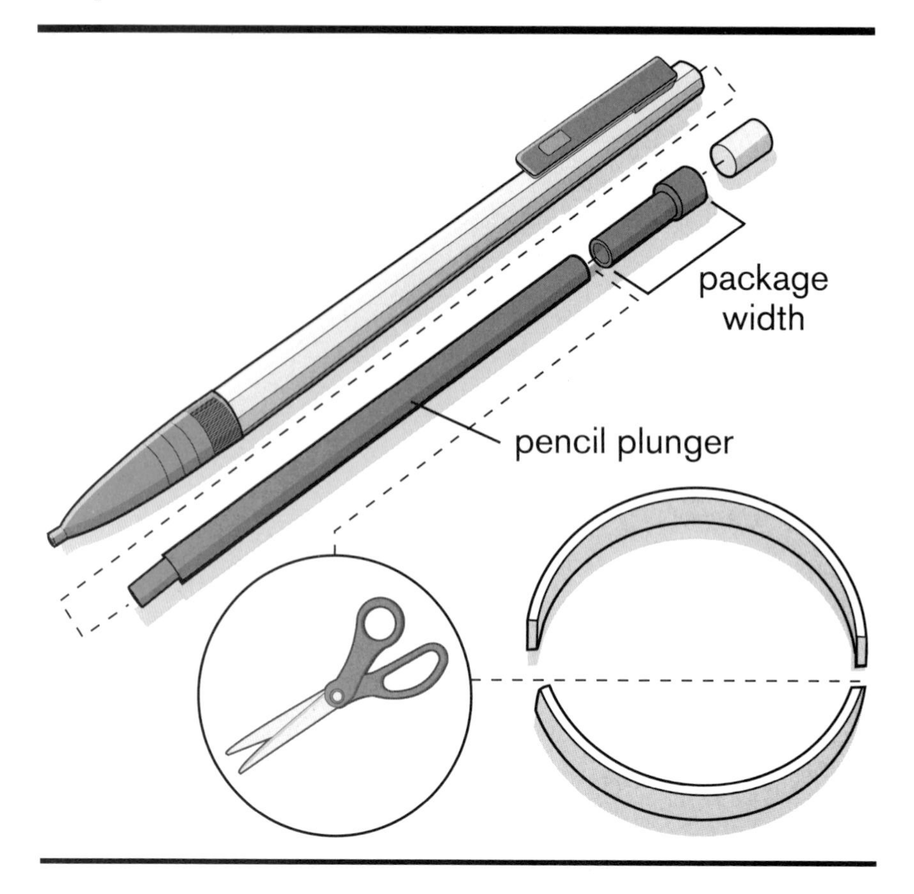

First, disassemble an inexpensive mechanical pencil. Pull out the pencil plunger and remove the eraser from the end.

Line up the width (side) of the Tic Tac container to the pencil plunger end, then remove that width from the end of the pencil plunger using scissors or a hobby knife, as shown. Discard the pencil and the full-length plunger scrap.

Finally, cut a wide rubber band in half as shown.

Step 2

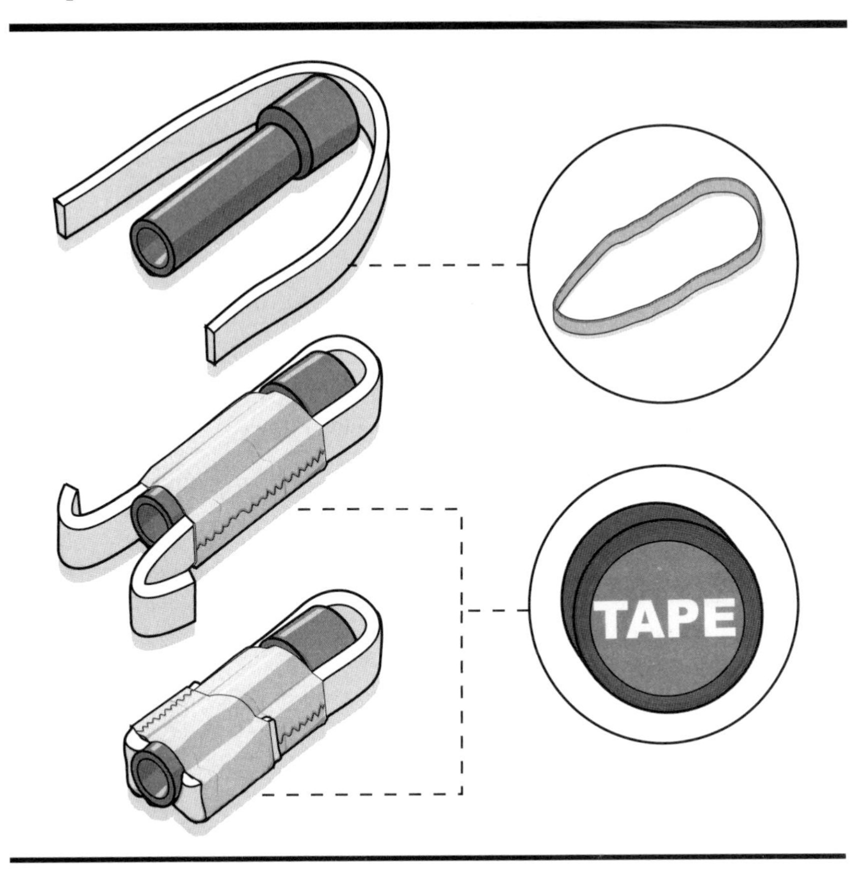

Cover the eraser end of the pencil plunger with the center of the rubber band half. Duct tape the rubber band onto the pencil plunger as shown; the rubber band should not be tight to the eraser end, as this will make grabbing it difficult.

Any extra rubber band should be overlapped to increase the strength of the hold (see the bottom images) and taped into place. If there is excess rubber band after you have folded it back and taped it, trim it off with scissors (not shown). The pencil plunger firing assembly is complete.

Step 3

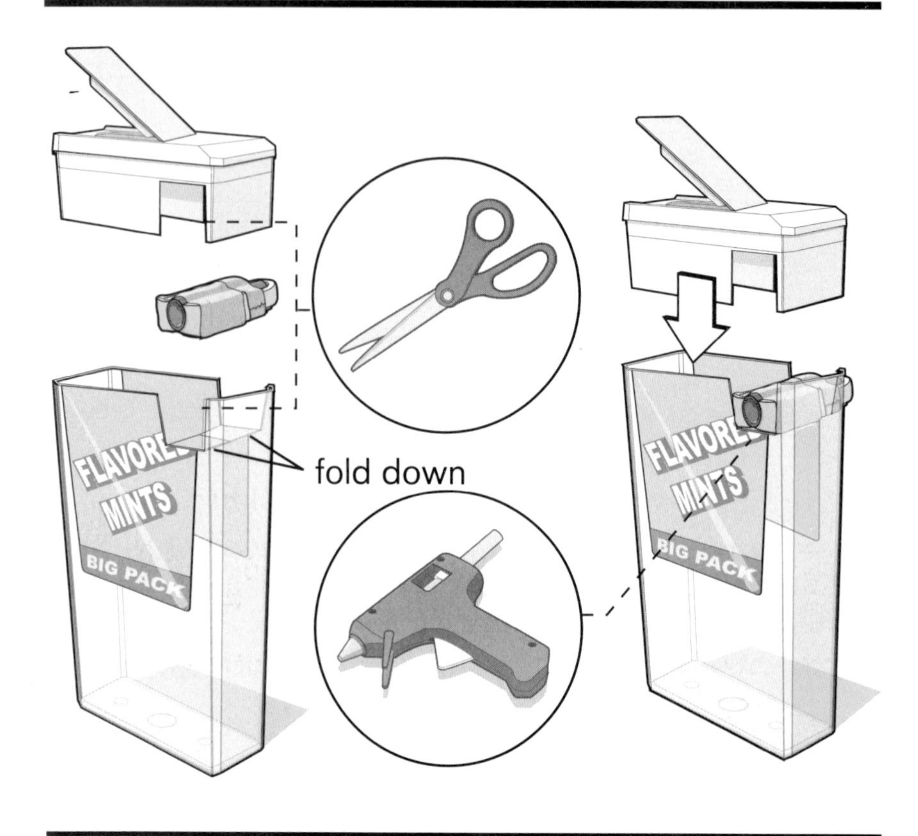

Remove the cap from the Tic Tac container. Use scissors to make four small slits in the container that are the approximate width and height of the pencil plunger firing assembly. Fold down the cut flaps toward each other, making a shelf on which to rest your firing assembly. (If the flaps are too brittle to fold, snap them off.)

Opposite the Tic Tac container "door," cut the same-sized opening into the cap as shown. Remove the material from the cap so it does not interfere. Hot glue the pencil plunger firing assembly onto the shelf inside the Tic Tac container. Then snap the cap onto the container.

To fire your Mint Pack and Arrow, remove one end from a cotton swab and load it into the pencil plunger opening, remaining swab side out. Find the cut end of the cotton swab in the rubber band, and then pull back the rubber band and release. The Mint Pack and Arrow can also fire toothpicks and modified pen ink cartridges. When using the weapon to fire toothpicks, remove the pointed end to avoid wear on the rubber band. And never aim this devious little launcher at anyone!

3

VILLAIN MINIWEAPONS

Q-PICK BLOWGUN

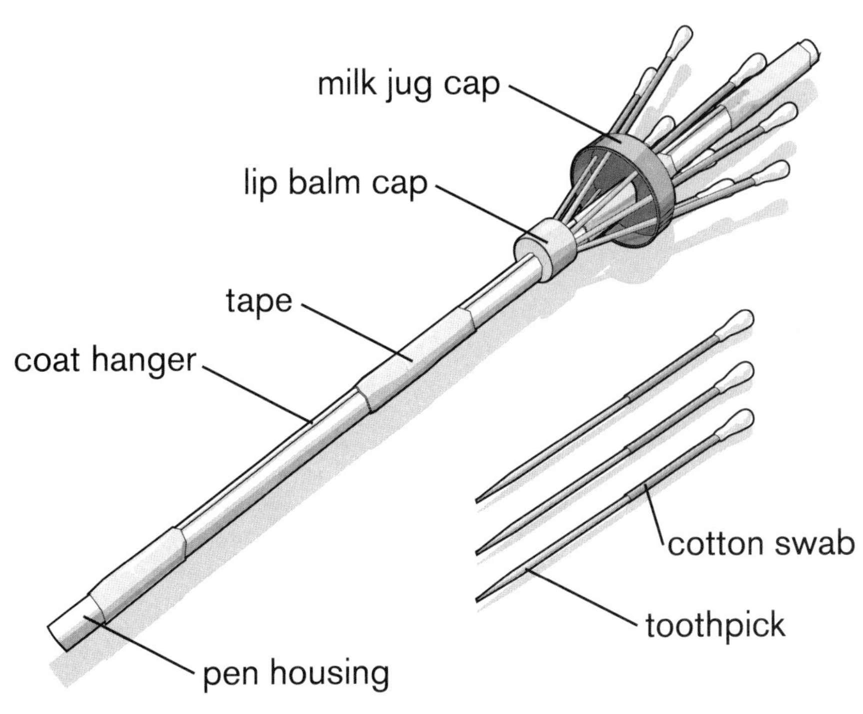

Range: 10–30 feet

With competition styling and standardized construction, the Q-pick Blowgun is one of the few tournament-graded MiniWeapons the department has developed. The plastic blowpipe and wooden-tipped darts can be built for less than pennies, making it an excellent choice for a small army of ninja assassins.

Supplies

3 plastic ballpoint pens
Duct tape
1 metal coat hanger
1 plastic milk jug cap (or similar)
1 plastic lip balm cap (or similar)

Tools

Safety glasses
Hobby knife or pliers
Wire cutter or pliers
Scissors
Hot glue gun (optional)

Ammo

1+ toothpicks
1+ cotton swabs (plastic)

Step 1

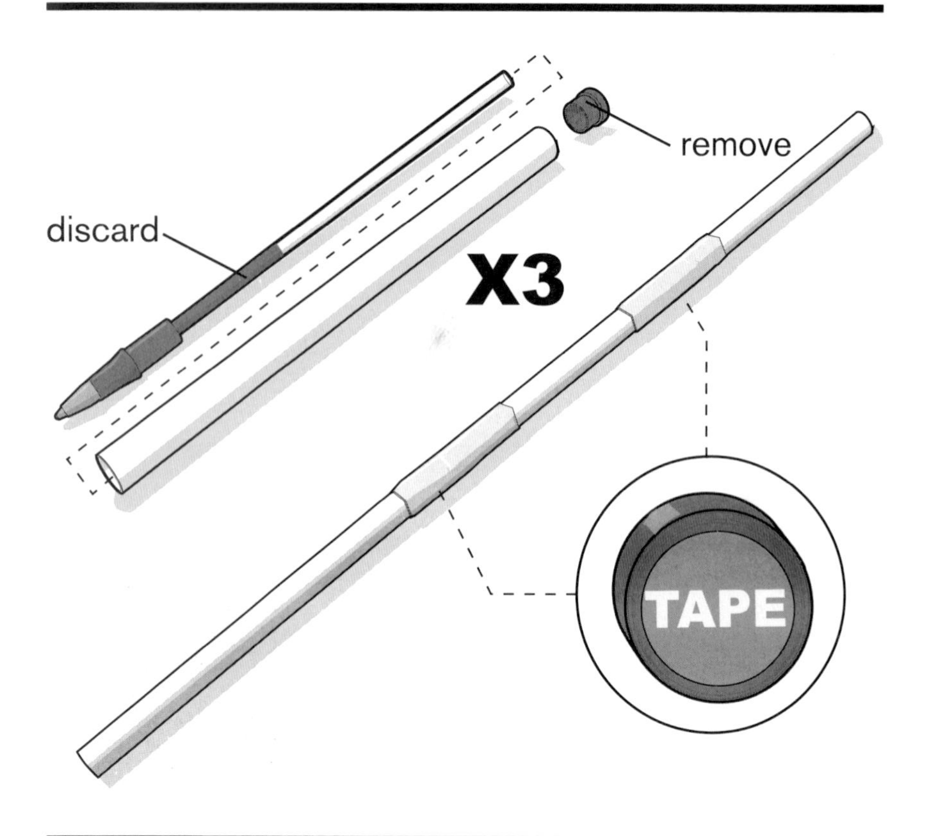

Search your junk or desk drawer for three plastic ballpoint pens, all similar in dimension and diameter. Disassemble each pen into its various parts. Depending on the pens, you may need a tool to dislodge the rear pen caps. A hobby knife or small pliers should be sufficient. Save the rear pen caps for building the Round-Nosed "Bullet" (page 73); the ink cartridge can be discarded.

Carefully duct tape all three pen housings together to create a long tube. The pen cylinders' alignment will need to be exact, so take your time and do a careful, thorough job. Inspect the cylinder by peering through it.

Step 2

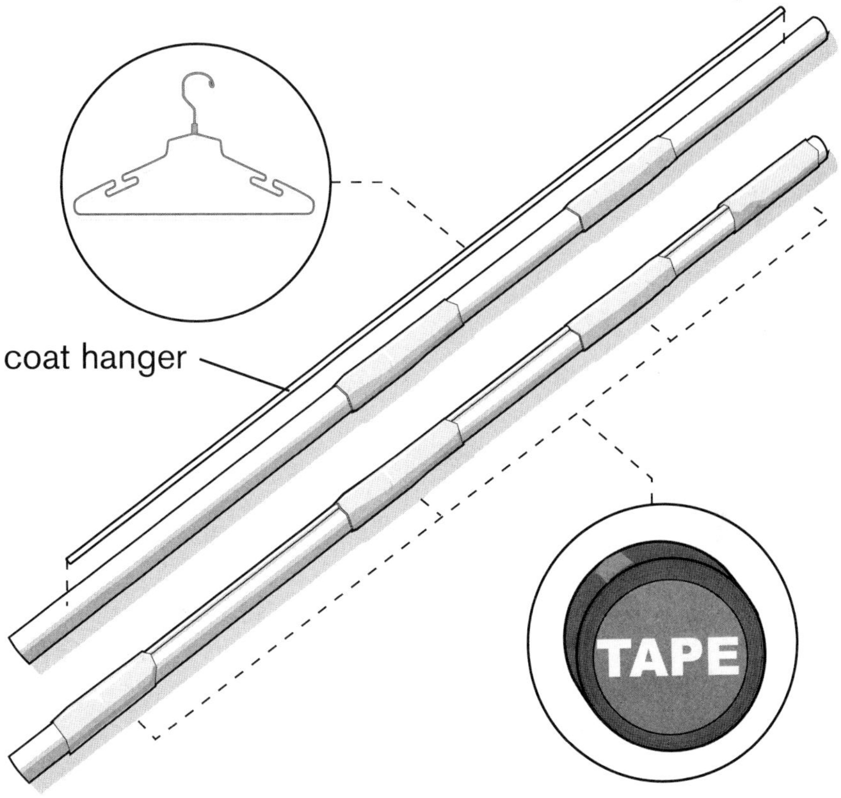

With pliers or wire cutter, remove the straight section from the bottom edge of a metal coat hanger.

Tape this section to the long cylinder of pen housings to strengthen the blowgun.

Step 3

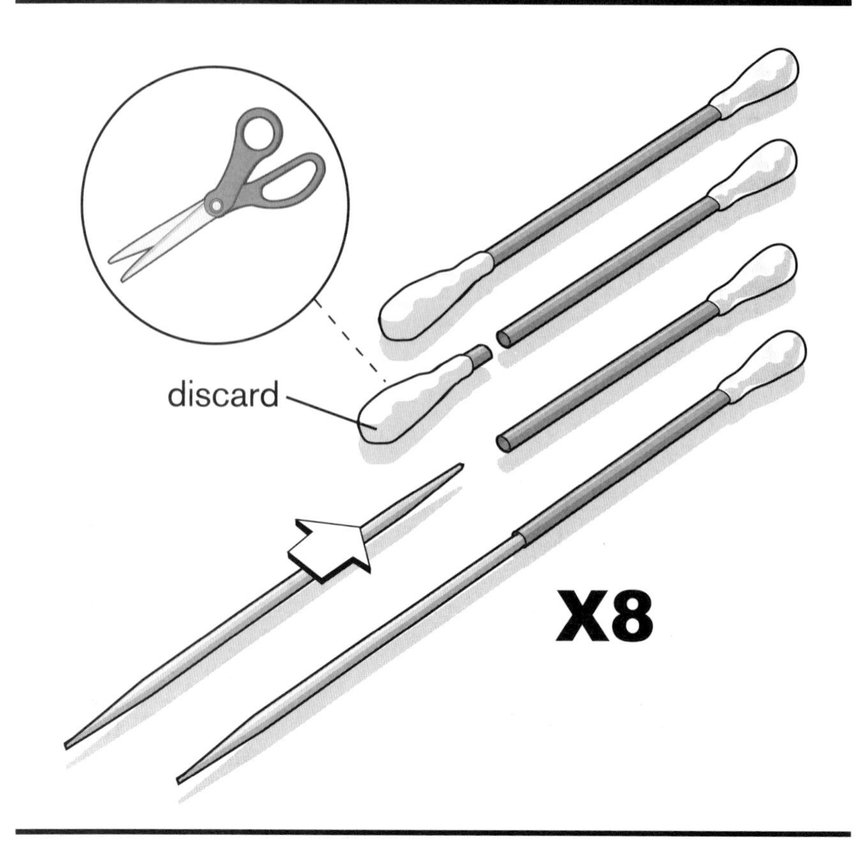

Each blow dart will be constructed from one plastic cotton swab and one toothpick.

Cut one end off the plastic swab as shown, then slide a single round wooden toothpick into the plastic cavity of the cotton swab. Push it in until it's tight. You can glue the pieces together, but it's probably not necessary. Repeat this step to create at least eight blow darts.

Step 4

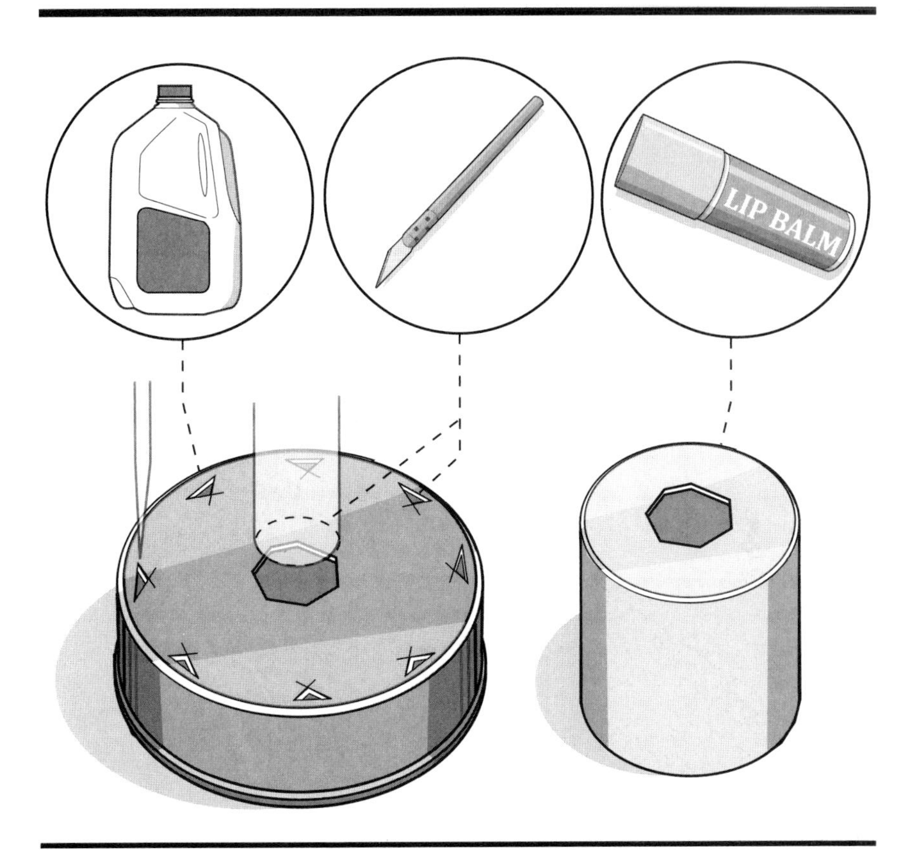

Now build the dart holder out of two plastic caps, the first being a milk jug cap and the second being a lip balm cap, or similar-sized cap. These two caps are suggested because they are manufactured using thin, soft plastic, making them easier to cut. (This dart holder is optional and will not affect the Q-pick Blowgun's performance.)

Using the pen housing as a template, cut two holes similar in diameter, each in the center of both the milk jug cap and lip balm cap using a hobby knife. The holes do not have to be perfectly round; you are only looking for a tight fit when they eventually slide onto the pen housing.

Next, cut eight evenly spaced smaller holes around the perimeter of the milk jug cap. Each hole should be roughly the same size as a the diameter of a wooden toothpick. Again, these holes do not have to be perfectly round—small triangular cuts might be easiest.

Step 5

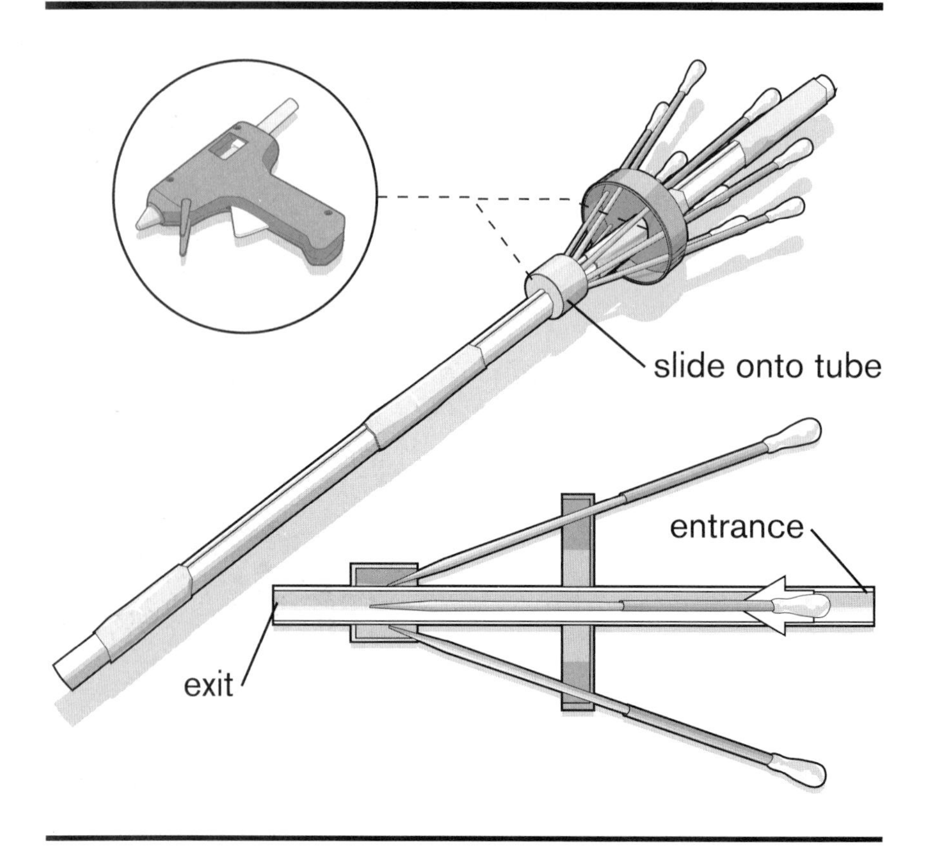

Slide the lip balm cap approximately 7 inches down the pen barrel. The flat, flush end of the cap should be facing the exit (long end) of the barrel, as shown. If the cap is loose, fasten it to the barrel with hot glue or tape. Next, slide the milk jug cap approximately 5 inches down the pen barrel, with the flush end of the cap facing the mouth end (dart entrance) of the barrel. Secure this cap in place using hot glue or tape.

Load the extra darts into the milk jug cap and angle them inward, into the lip balm cap. This should hold the darts in place; adjust or modify the spacing if needed.

Now stuff a Q-pick dart into the pen housing with the toothpick point facing toward the barrel exit. Take a deep breath, aim, and exhale sharply to blow out the dart. ***Never inhale while the pen housing is in your mouth waiting to be launched.*** Always be responsible and fire your Q-pick Blowgun in a controlled manor, never at any living thing. Safety glasses are a must, as is staying clear of spectators.

RUBBER BAND DERRINGER

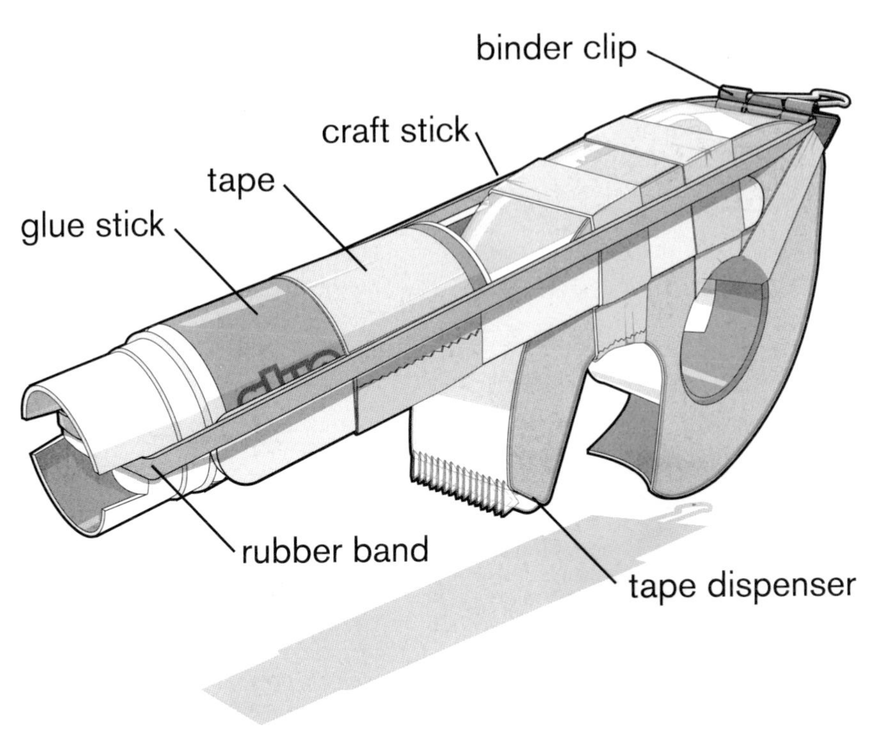

Range: 8–20 feet

Villains are crazy for this single-shot Rubber Band Derringer because it can be operated with one hand, freeing the other hand to clutch stolen jewels or loot! Its design is easily concealable, and loading the elastic bullets is quick and easy. Release the binder clip hammer and watch the rubber band zing off toward its intended target. But be on alert–other owners of this MiniWeapon often sport double derringers!

Supplies

1 glue stick container
Duct tape
2 craft sticks
1 plastic tape dispenser
1 small binder clip (19 mm)

Tools

Safety glasses
Pliers
Hobby knife

Ammo

1+ rubber bands

Step 1

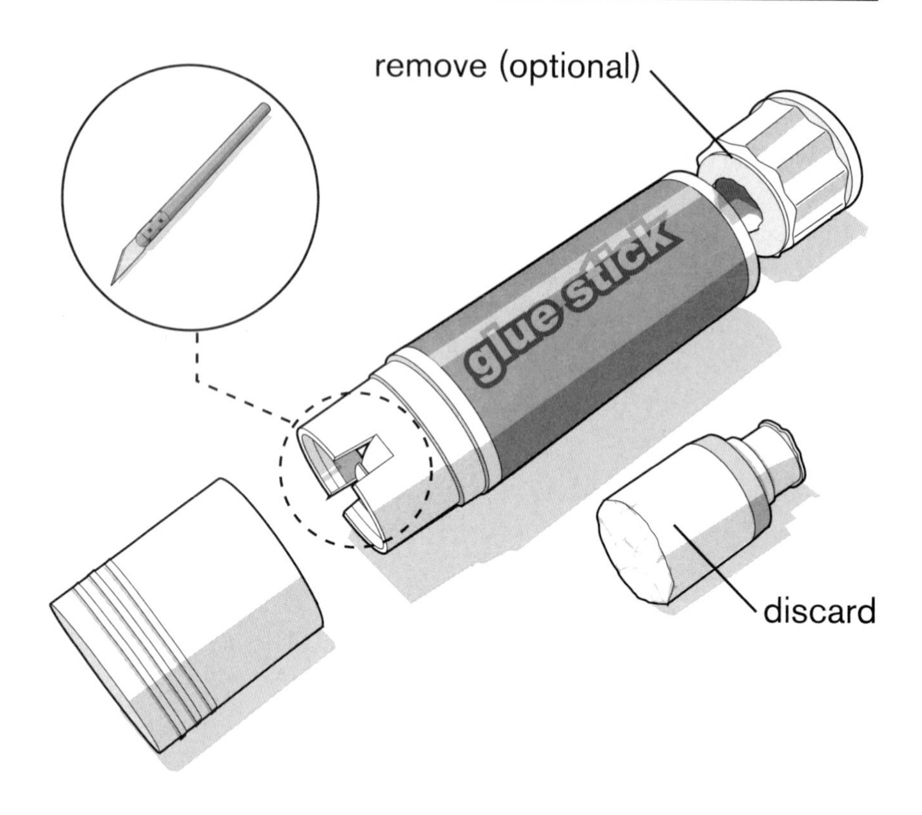

Your first step is to disassemble a glue stick or similar small-cylinder container. To remove the insides, snap off the bottom twist. You may need pliers to assist you. Once the bottom twist is off, remove the rotating bar and any remaining glue inside. Discard the insides and bottom twist; however, save the cap for the Q-pick Blowgun (page 111).

On the open side of the small cylinder, use a hobby knife to cut two small slits on each side, about ¼ inch apart. Then bend the plastic flaps between the two cuts inward, toward each other. This groove will hold the rubber band in place.

Step 2

Using duct tape, fasten two craft sticks parallel to each other on the opposite sides of the small container. The craft sticks should align with the small slits you cut in the first step (glue stick opening), but at the other end of the tube. See the illustration. This completes the simple barrel assembly.

Step 3

Now tape the barrel assembly to an upside-down plastic tape dispenser. This dispenser will be a quick and easy gun stock and handle, complete with a makeshift trigger guard.

Step 4

Use duct tape to attach the small binder clip to the back of the tape dispenser. To test the clip's location, load a rubber band through the clip with the other end around the tip of the barrel, securing it into the two cut slots at the end. While wearing safety glasses, carefully aim the launcher away from you and press down on the top metal handle. If the rubber band does not launch, readjust the clip as needed (probably forward).

Never aim your Rubber Band Derringer at a human or animal, but defiantly at a robot! ***A rubber band can sting or cause eye damage, so always wear safety glasses when firing.***

Alternate Construction

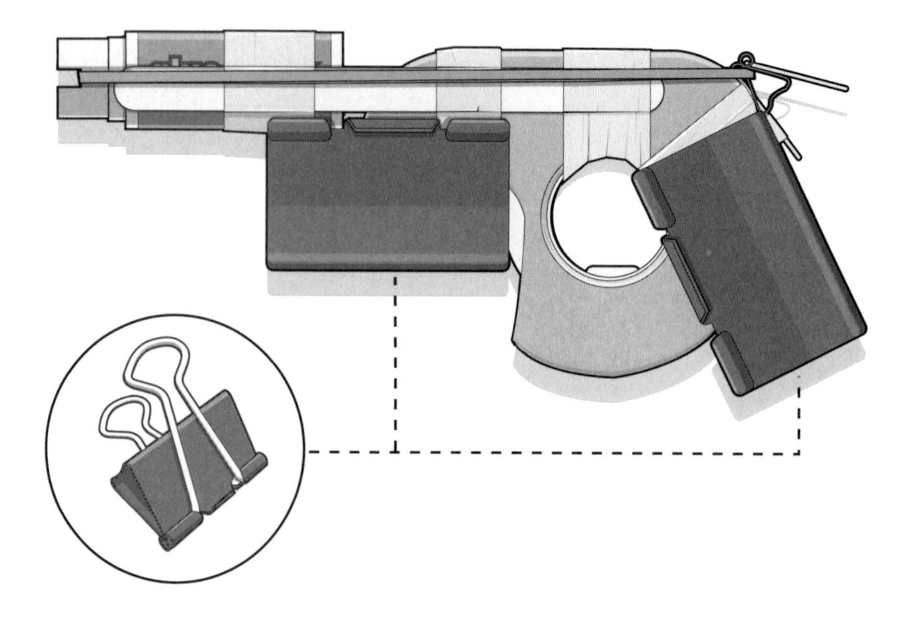

The Rubber Band Derringer's generic design invites custom modifications. If you want to add some bulk to your shooter, clip a few large binder clips (51 mm) to the underside of the barrel and handle, as shown. Additional upgrades can be made using various paper tubes around the barrel and other cylinders. Have fun with it!

DOUBLE-BARRELED BAND GUN

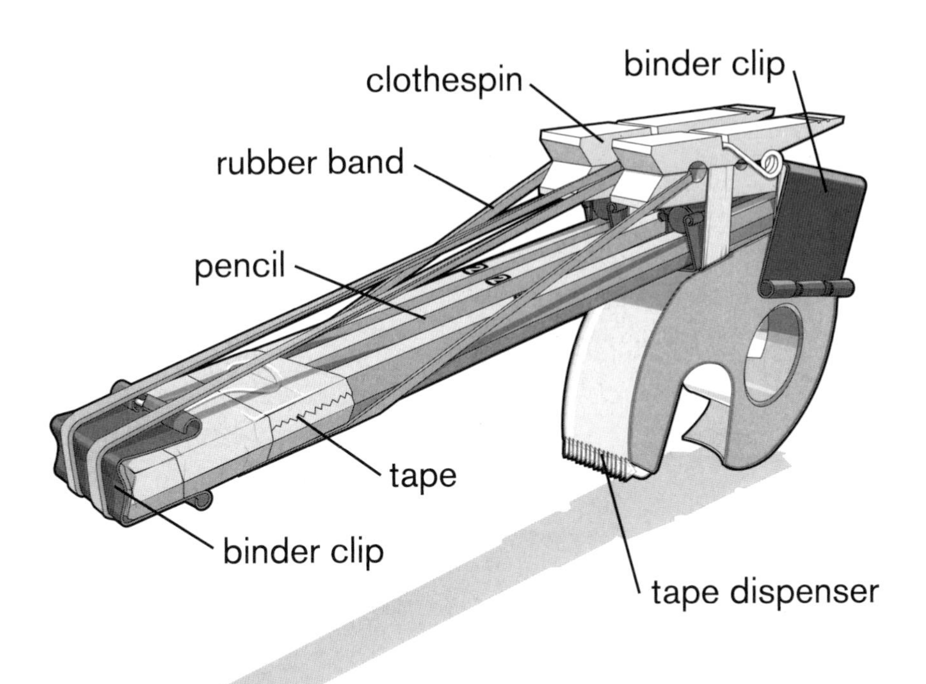

Range: 8–20 feet

When not holstered, the Double-Barreled Band Gun packs a one-two punch. When fired, the elastic bands are silent, avoiding the sonic boom caused by some of the bean shooter–type sidearms found in chapter 1. Designed to fire each rubber band individually, this gun also allows you to unload both barrels simultaneously—the perfect indoor MiniWeapon!

Supplies

3 small binder clips (19 mm)
3 pencils
Duct tape
2 wooden clothespins
1 large binder clip (51 mm)
1 plastic tape dispenser

Tools

Safety glasses

Ammo

2+ rubber bands

Step 1

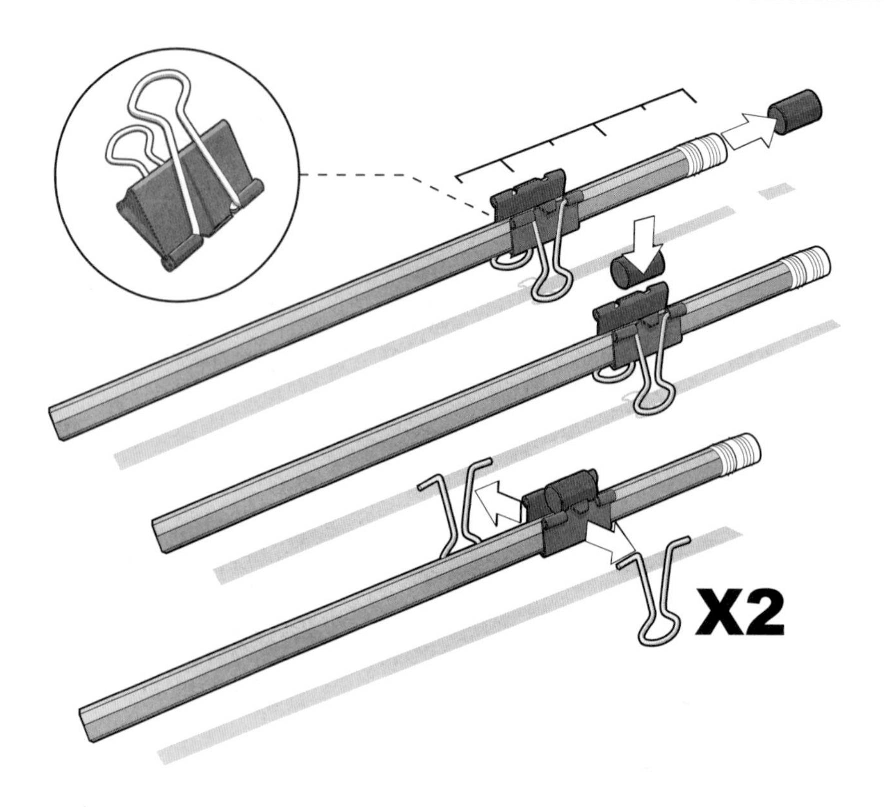

Fasten a small binder clip 2½ inches from the eraser end of a pencil. Then gently dislodge the eraser from its metal casing by twisting or rocking it. Place the removed eraser into the clip as shown. Once the eraser is in place, remove the metal handles from the clip.

Repeat this step one more time, for two complete "barrels."

Step 2

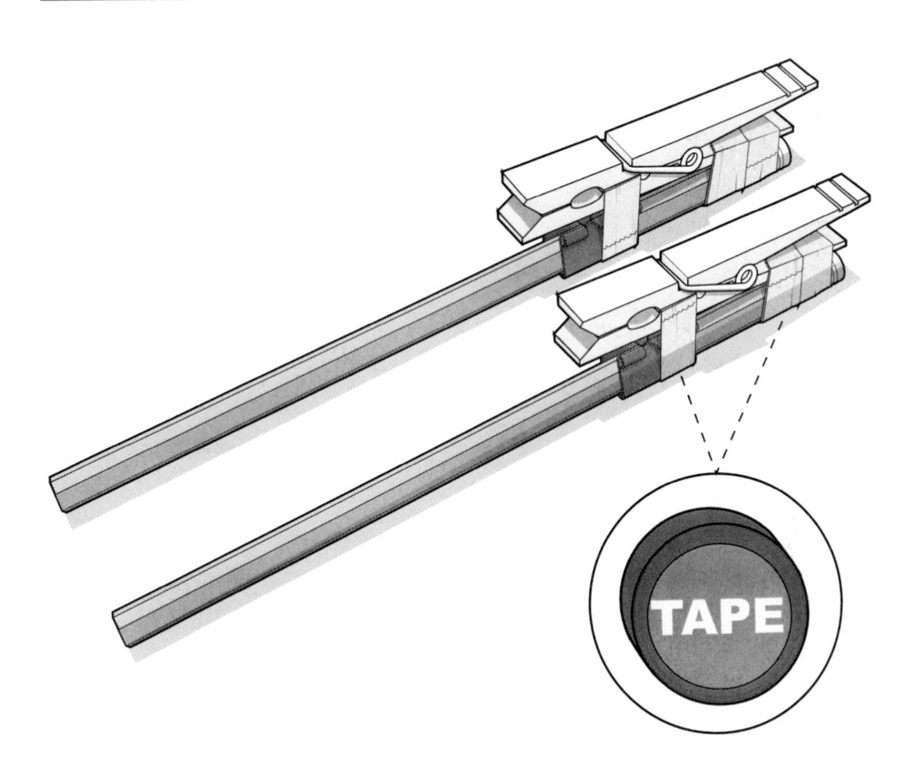

Now duct tape a wooden clothespin to the top of one binder clip/pencil assembly. The duct tape should be exceptionally tight and thorough; a loose fit will cause the assembly to slide forward when loaded with a rubber band bullet—not good.

Repeat this step one more time, for two complete builds.

Step 3

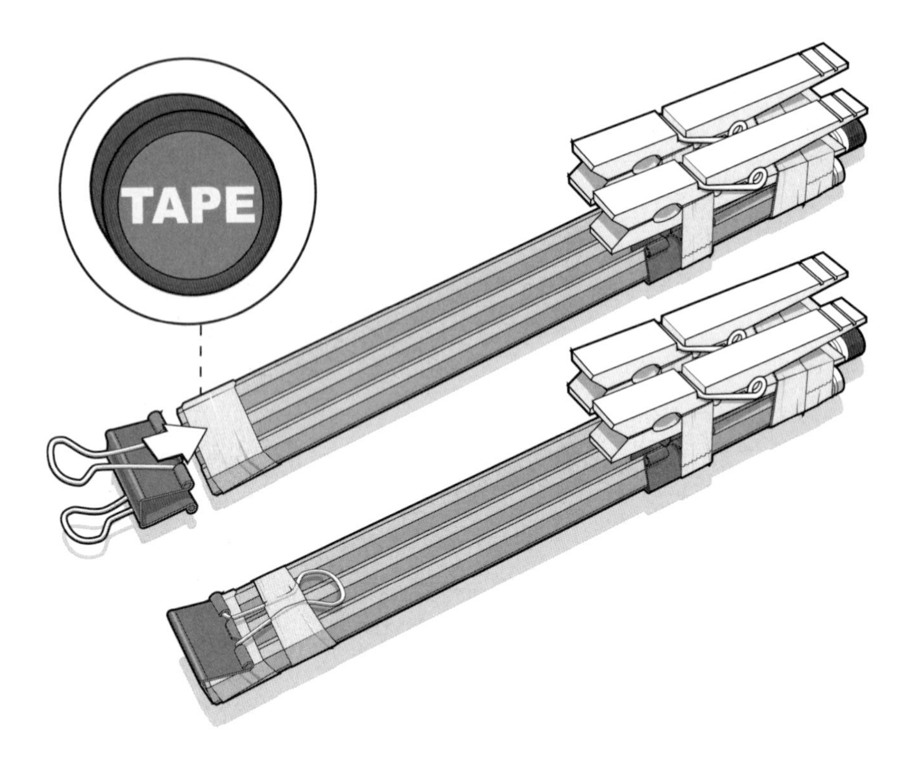

Place a third pencil between the two modified pencils fixed with clothespins and line up all three pencils so the ends are flush. Once in place, duct tape the pencil ends together to hold all three in place.

Now place a small binder clip over the end of the three pencil ends as shown. Fold the metal handles down and duct tape them against the pencil housing.

Step 4

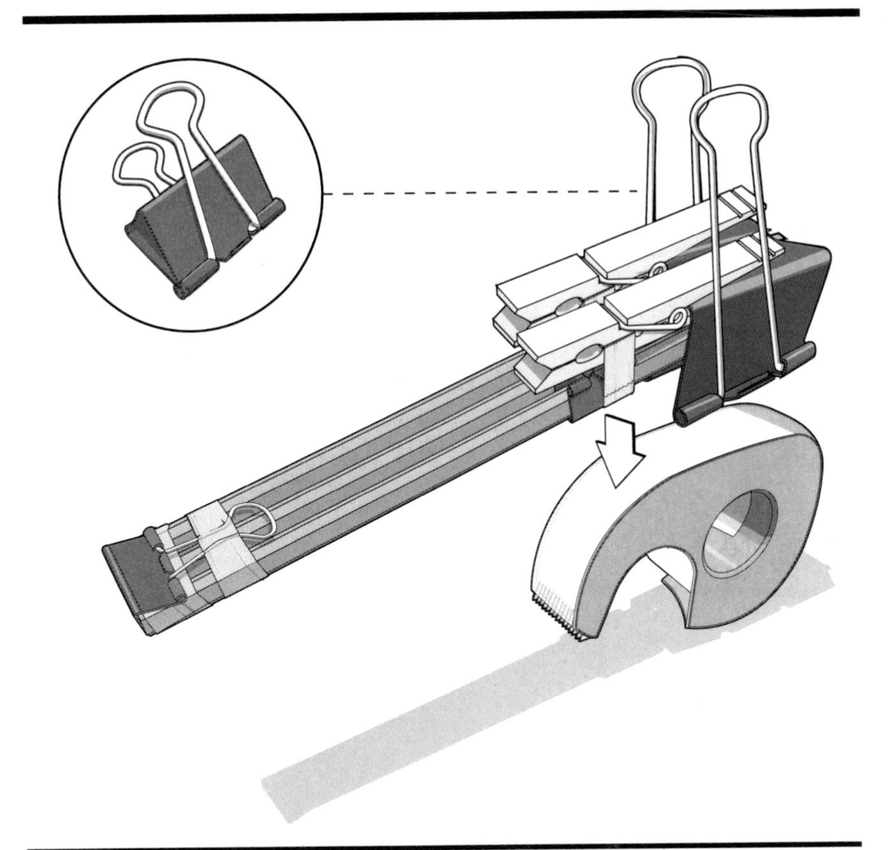

Slide a large binder clip over the back end of the pencil assembly as shown. Do not cover the top segment of the clothespins.

Snap the entire assembly onto the bottom surface of a plastic tape dispenser.

Step 5

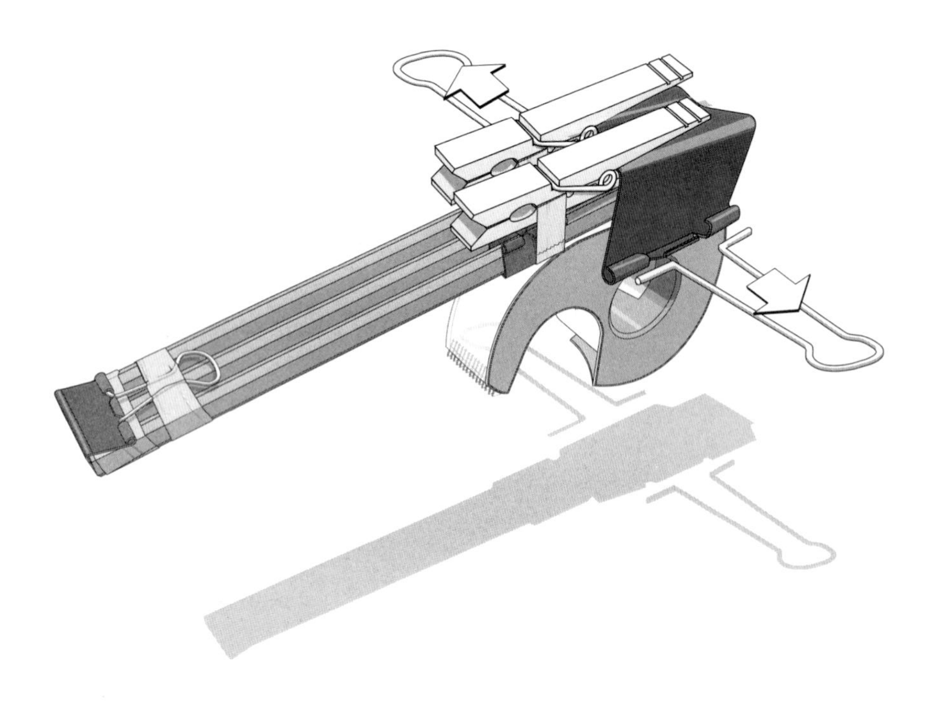

If you have a tight fit, remove both metal handles from the large binder clip. If your assembly seems loose, flip down both metal handles on the big binder clip so they rest on the side of the tape dispenser. Add duct tape if the fit is loose.

Now load each barrel with a rubber band. First clip the rubber band inside the clothespin, and then wrap the loose end of the band around the tip of the launcher. Repeat this step on the opposite side. Prior to launching, the rubber band should be fixed as shown in the first illustration (page 123).

When operating your Double-Barreled Band Gun, wear safety glasses and always prepare for the unexpected. ***Never aim your rubber band gun at a living target.*** Always use common sense, and use this weapon at your own risk.

PUSHPIN DART

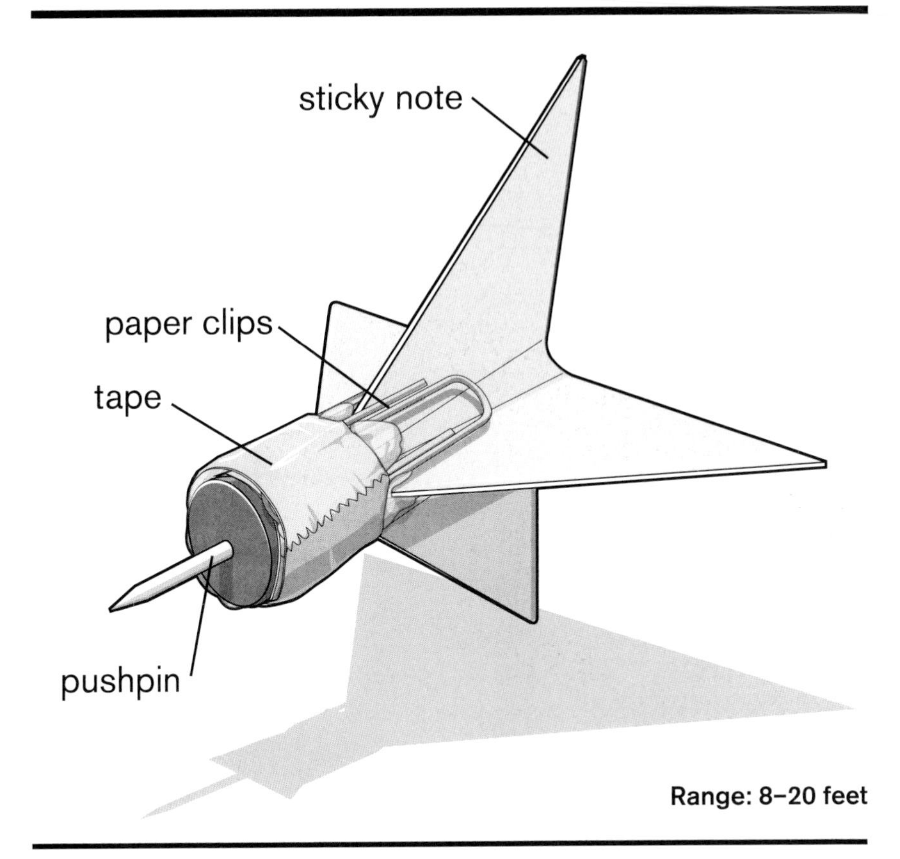

Pushpin Darts are awesome! Equipped with a factory-molded integrated metal point, each dart is extremely durable. Its simple shaft and flight design is constructed from paper clips and one sticky note, making it colorful and practically free to build!

Supplies

Duct tape
1 pushpin
4 small paper clips
1 square sticky note (3 inches by 3 inches)

Tools

Safety glasses
Hot glue gun
Scissors

Step 1

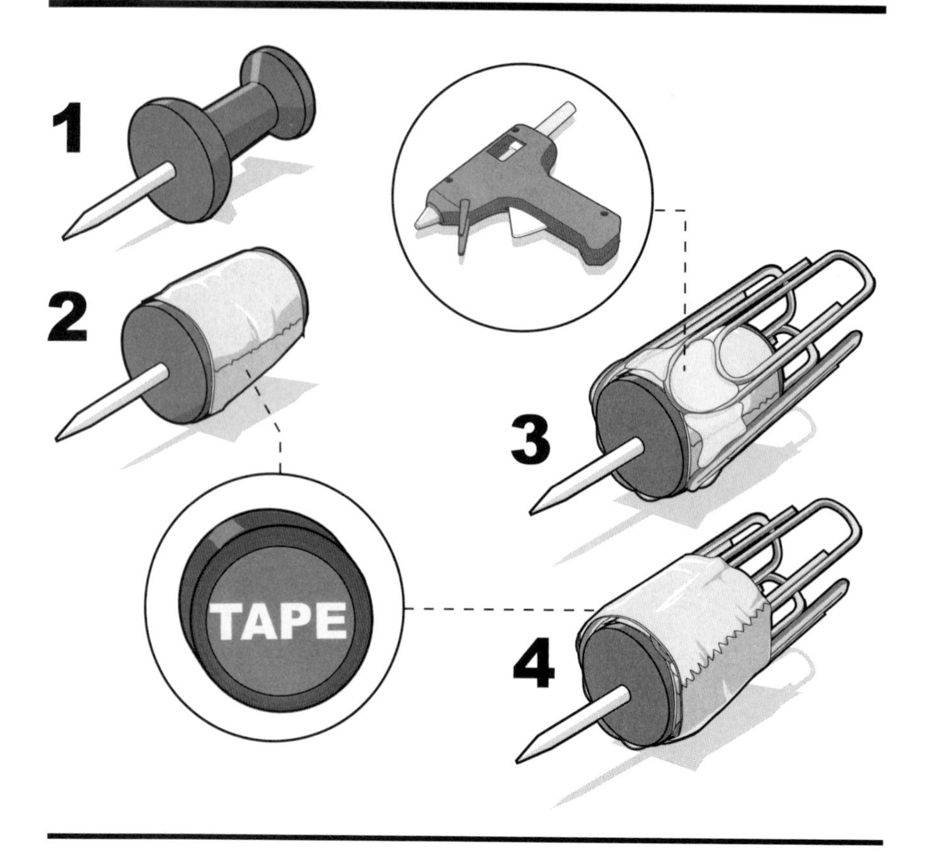

Wrap tape around the handle of a pushpin (1) as shown (2). Build up the pin's diameter so it's even with the rear plastic push.

Next, evenly space four small paper clips around the newly wrapped tape and hot glue them in place (3). After the glue has cooled, cover the glued paper clips with tape (4).

Step 2

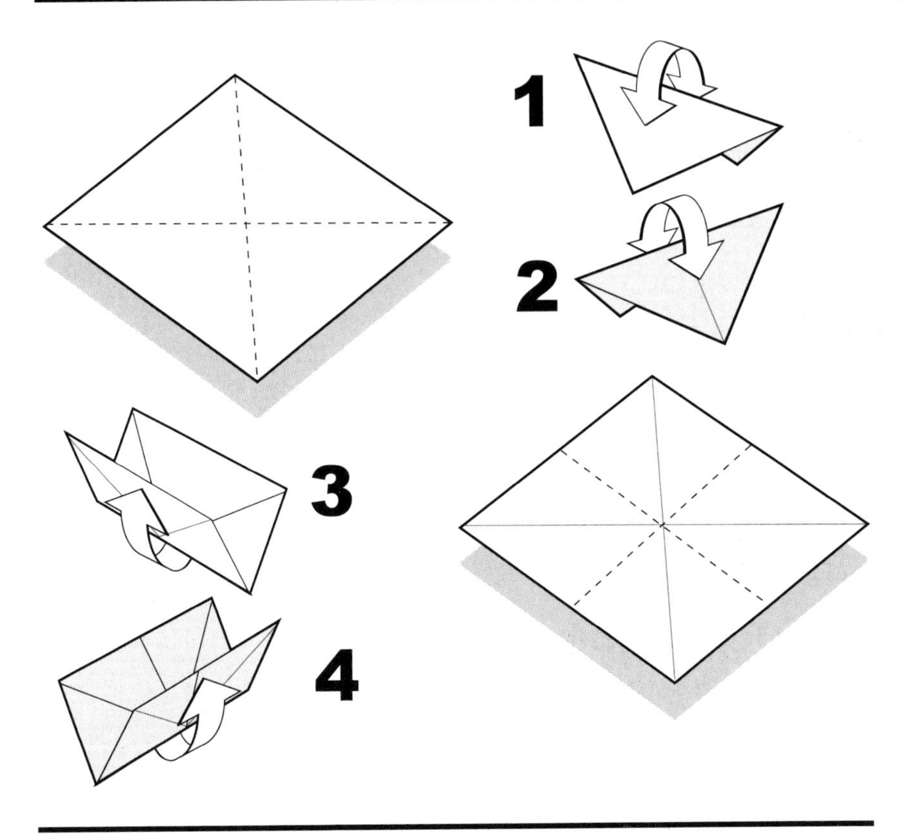

Fold the opposite corners of the sticky note together to form two triangles (1), then unfold the paper and fold the other two corners together to form two triangles, as shown (2).

Open the sticky note flat, then fold the paper in half both ways (3 and 4). Once you have finished, you should have folded the paper a total of four times and should have four crease lines, as shown above. These crease lines will act as guides for the major folds in the next step.

Step 3

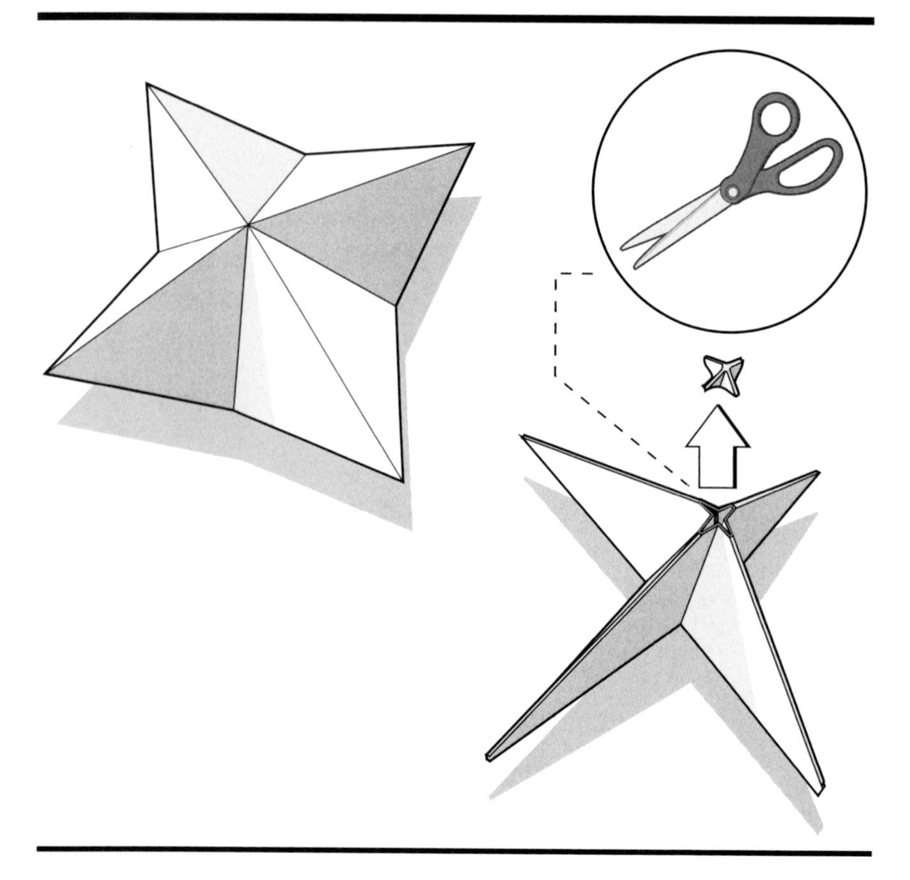

Now, creatively fold the square to create the flight wings or fins. Push up the center point while pushing down on the middle of the sides until you have what appears to be a star. This may take a few attempts due to the confusing nature of the creases and the stubbornness of the paper. Use the illustration above for reference.

Once you've completed the folds, run your fingers along the fold lines to sharply crease the edges. They will help the paper keep its form during construction.

Then use scissors to cut off the very front tip of the newly folded fins.

Step 4

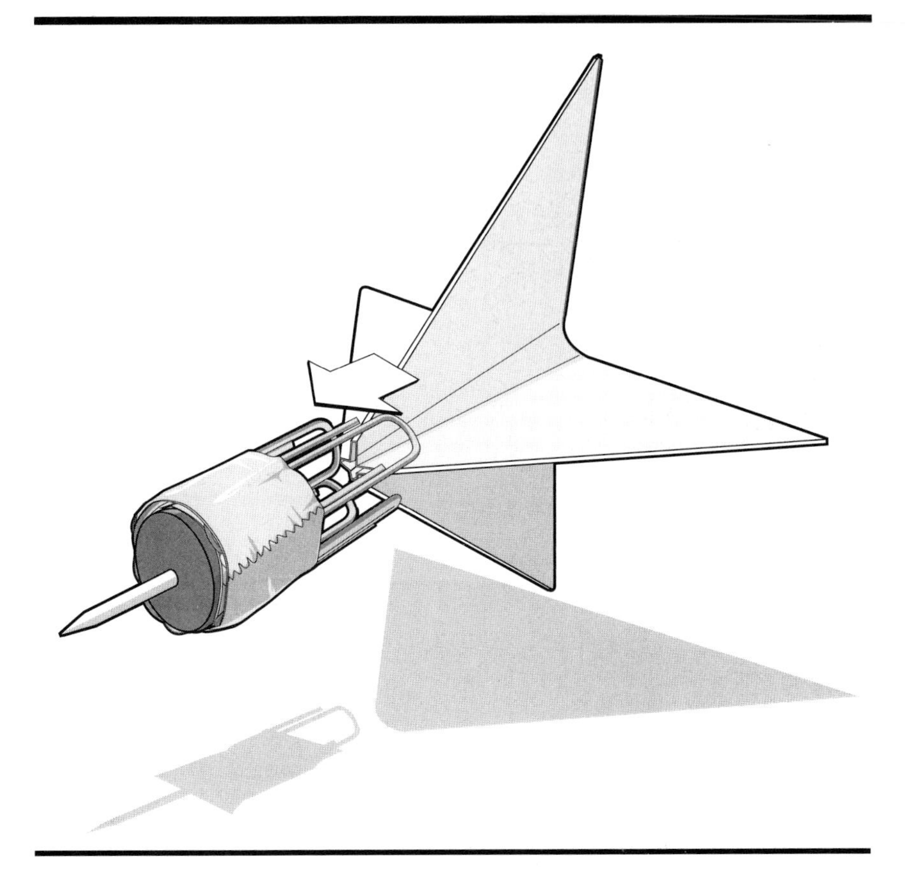

Slide the folded fins between the four paper clips. You may need scissors to remove a bit more paper from the tip to ensure a tight fit.

Step 5

Add hot glue around the areas where the paper clips and paper meet. This will ensure the fins don't fly off when shooting a friendly game of darts. If you need to practice your throwing, check out the printout targets in chapter 7 (page 249).

BOWLER HAT LAUNCHER

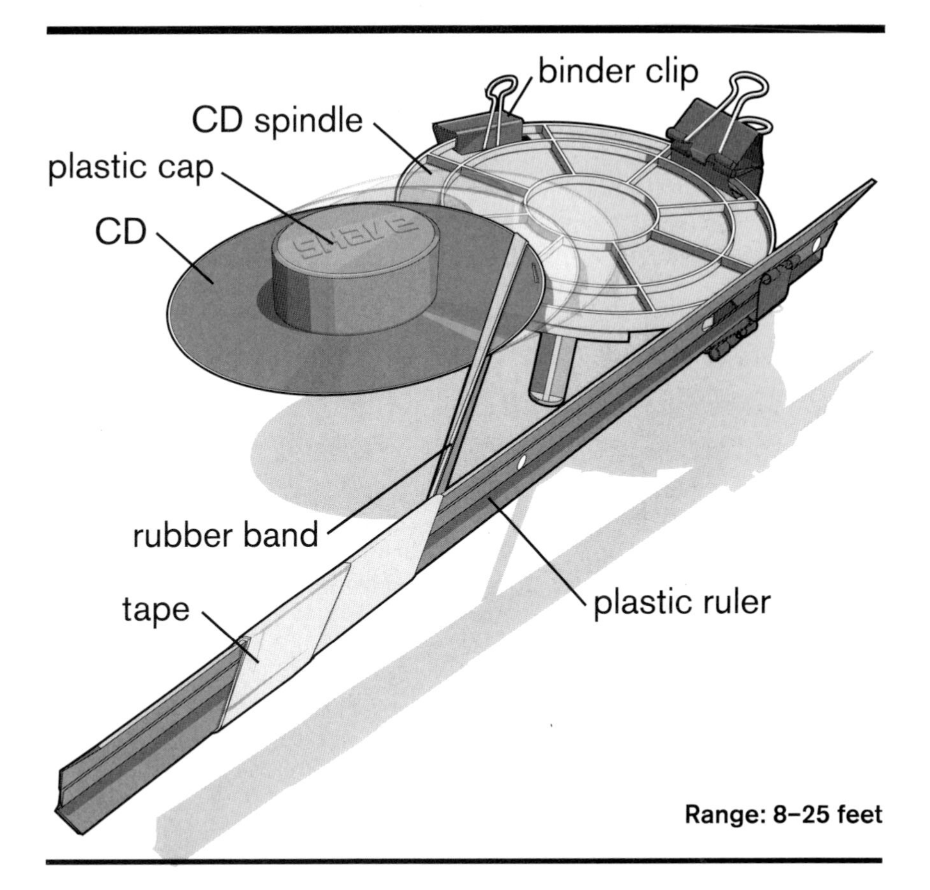

Even if a villain is superior in hand-to-hand combat, he or she might rely on some "oddjob" weapons. The Bowler Hat Launcher ejects a lethal flying disc disguised as a gentleman's hat. The flying disc could also act as a clay pigeon launcher—if you wished to test your aim of a different MiniWeapon using a moving target.

Supplies

1 CD spindle bottom
5 medium binder clips (32 mm)
Duct tape
1 plastic ruler
1 wide rubber band

Tools

Safety glasses
Dremel or wood-burning tool
Pliers
Hot glue gun

Ammo

1+ unwanted CDs or DVDs
1+ plastic caps (approx. 2.5 inches in diameter)

Step 1

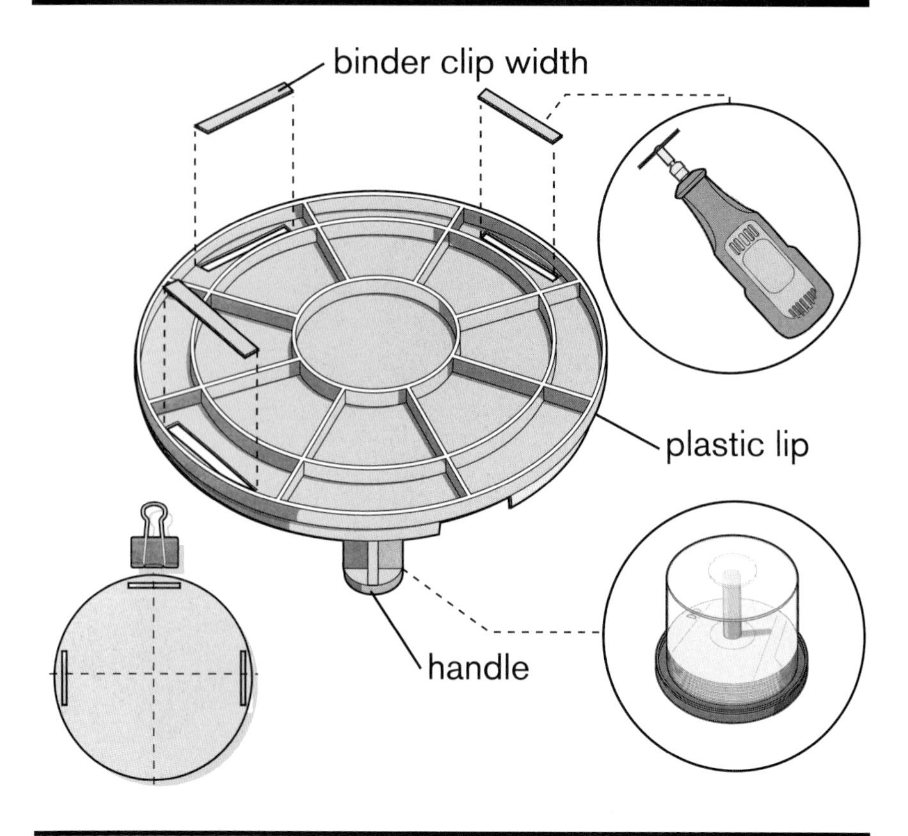

You will only be using the bottom of a CD spindle. (You can save the top of the spindle to make the CD-Spindle Catapult in the first *Mini-Weapons book*.) Flip over the disc spindle so that the protruding tower can function as a handle.

Depending on the configuration and thickness of the plastic, you might need a Dremel or wood-burning tool to complete the rest of this step. If the medium binder clips do not firmly clasp to the base of the spindle as shown, cut three holes in the surface of the CD spindle, each the length of the binder clip. The width of these holes needs to just allow one side of the binder clip to pass through.

All three holes will be made around the perimeter, two directly across from one another and the third between the first two. Use the illustration as reference. Again, if the CD spindle has a large plastic lip, you might not need to Dremel (cut) the housing.

Step 2

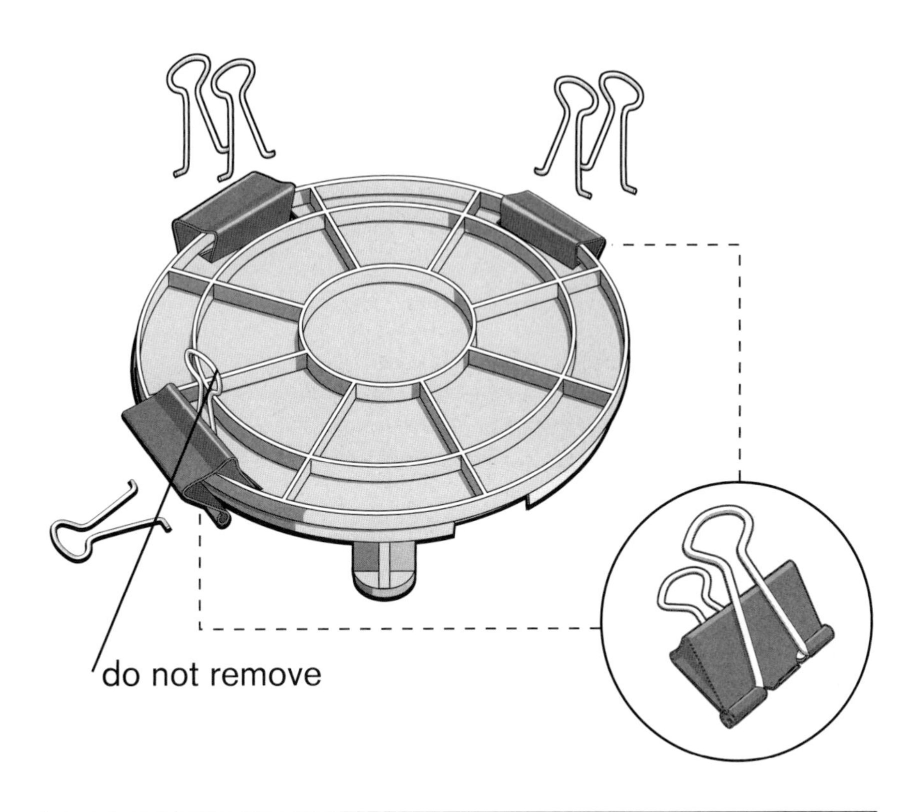

For each of the three binder clips, slide one end through the newly cut slots and clip the other end around the side of the CD spindle.

Remove all but one of the metal handles from the clips. You will keep only the left clip's inner handle in place (see above). This handle will be a safety guide for the disc.

Step 3

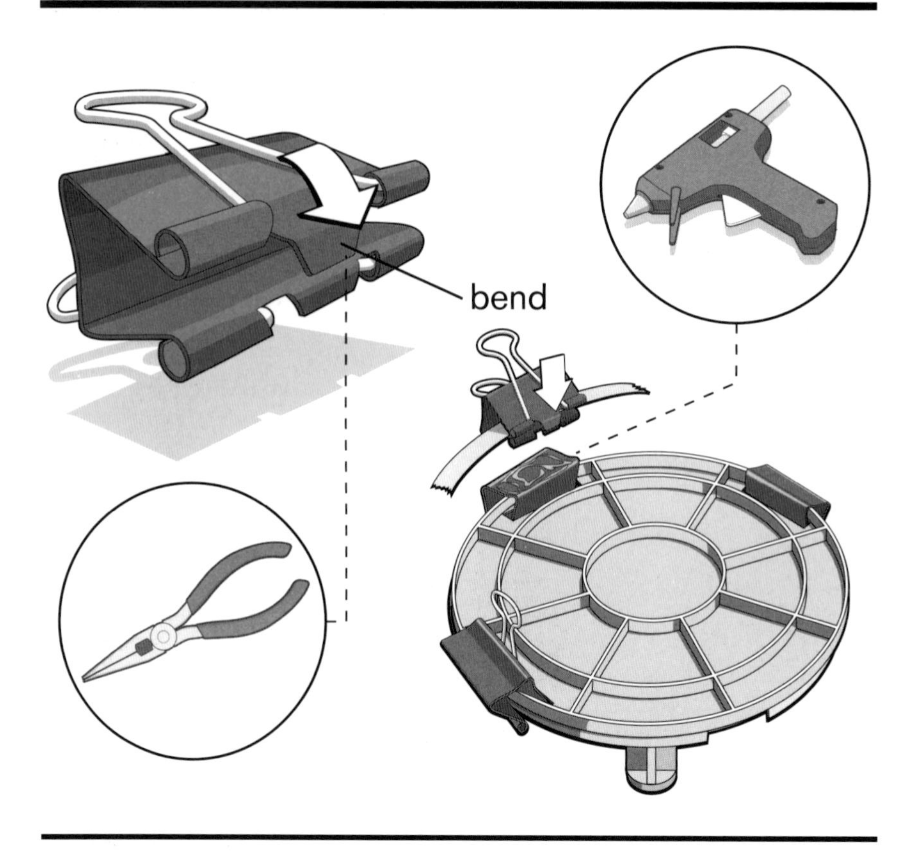

Now take another medium binder clip and, using pliers, bend the center middle tab 180 degrees, toward the mouth of the binder clip. This middle tab detail will clip into the disc and hold it in place before launching.

With the metal handles still attached, hot glue the binder clip to the center binder clip already fastened to the CD spindle, as shown. Add duct tape to the inside and outside of this clip to increase its holding power. It can't be too strong; this clip will be under a lot of pressure when the Bowler Hat Launcher is loaded.

Step 4

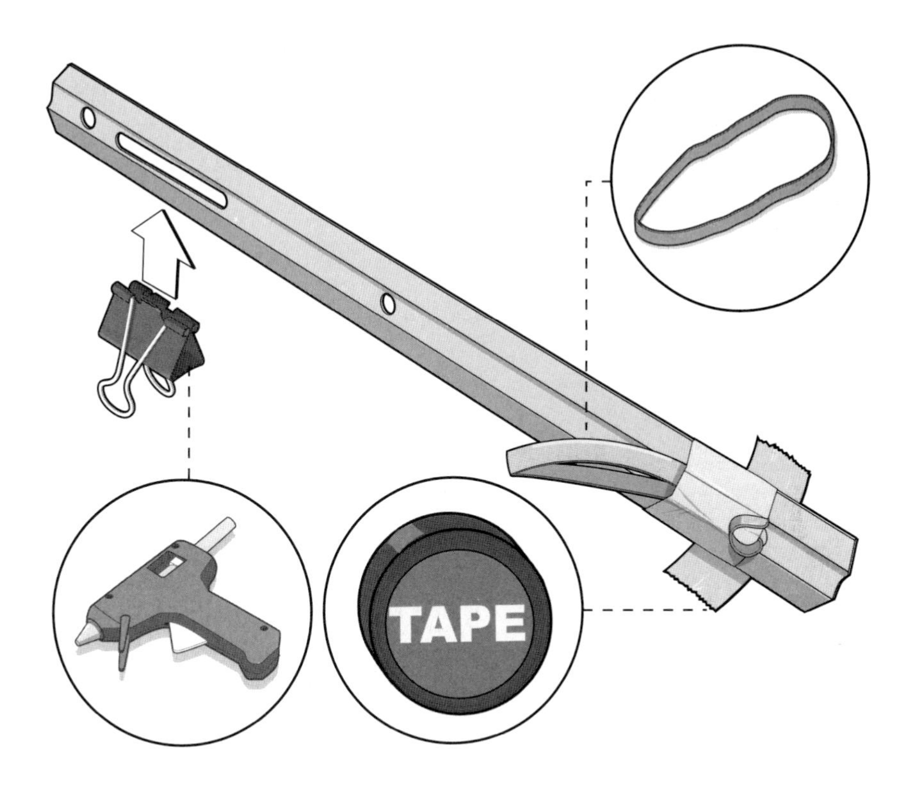

Now fasten the last binder clip to one end of the plastic ruler. Depending on the ruler's design, you can use the factory details molded into the ruler to increase the binder clip's holding power. Add hot glue if needed. Once snapped onto the ruler, remove the metal handles from the binder clip.

Next, tape one wide rubber band to the ruler at the opposite end of the fastened binder clip. Test its position by stretching it to the far end of the ruler. If you have a nice, tight feel with little ruler flex, your placement is perfect. Once you have finished this step, you still have the opportunity to easily adjust the rubber band for desired results.

Step 5

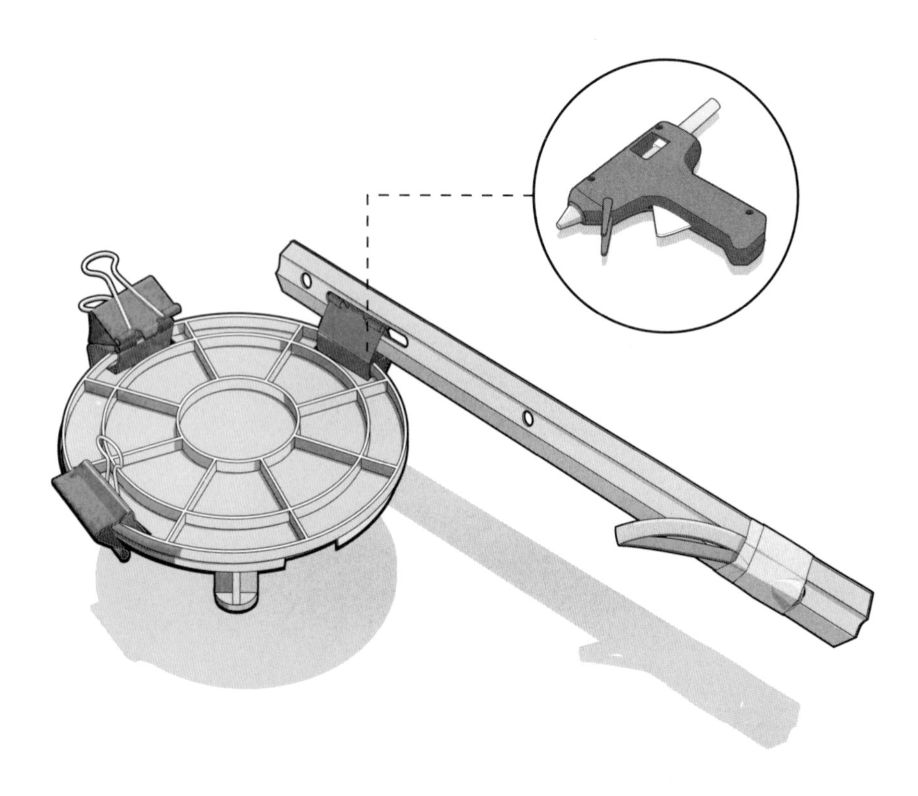

Load the top of the binder clip—opposite from the clip with a single handle hanging down—with hot glue, then carefully press and hold the ruler assembly clip to the hot glue. Hold the assembly together until the glue has cooled. If you are wary of the connection's strength, substitute superglue and add duct tape around this connection.

Step 6

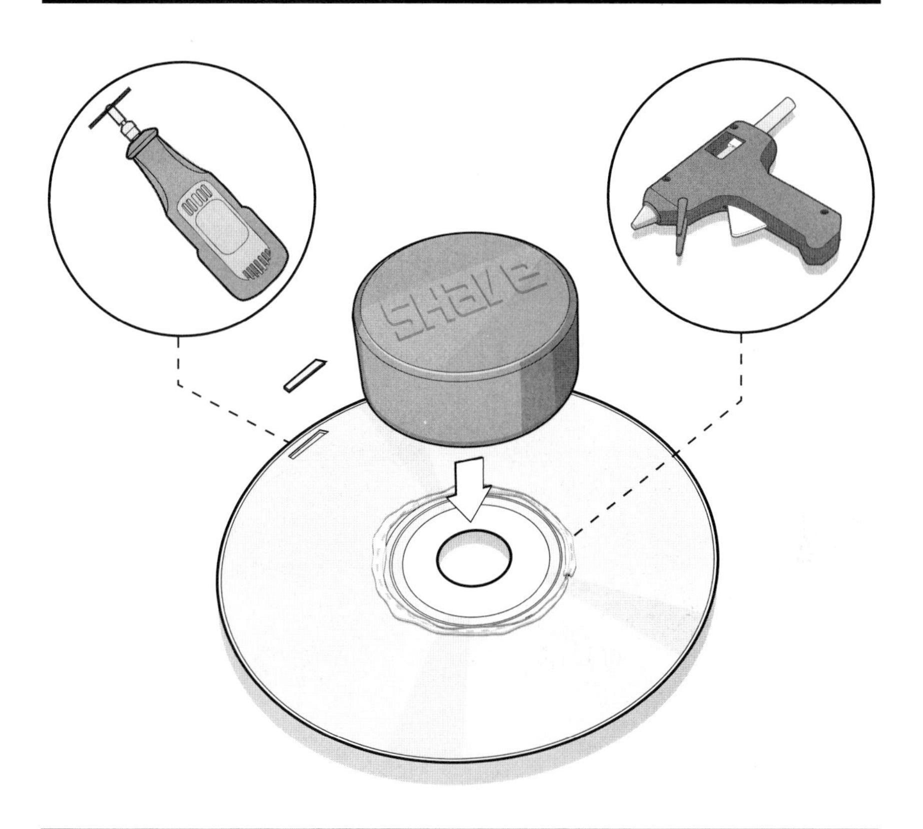

Using a hobby knife to cut a compact disc is very difficult, if not impossible. Instead, use a Dremel or wood-burning tool to cut a very small slit in the face of an unwanted CD or DVD. The slit just has to exceed the width of the bend you made in the medium binder clip (in step 3).

If you'd like the disc to resemble a bowler hat, like a famous super-villain's hat, hot glue a small plastic cap to the top of the disc. But remember that any additional weight to the disc will slow it down. Without the added plastic cap, this launcher is quite capable of flinging a disc over 30 feet.

Step 7

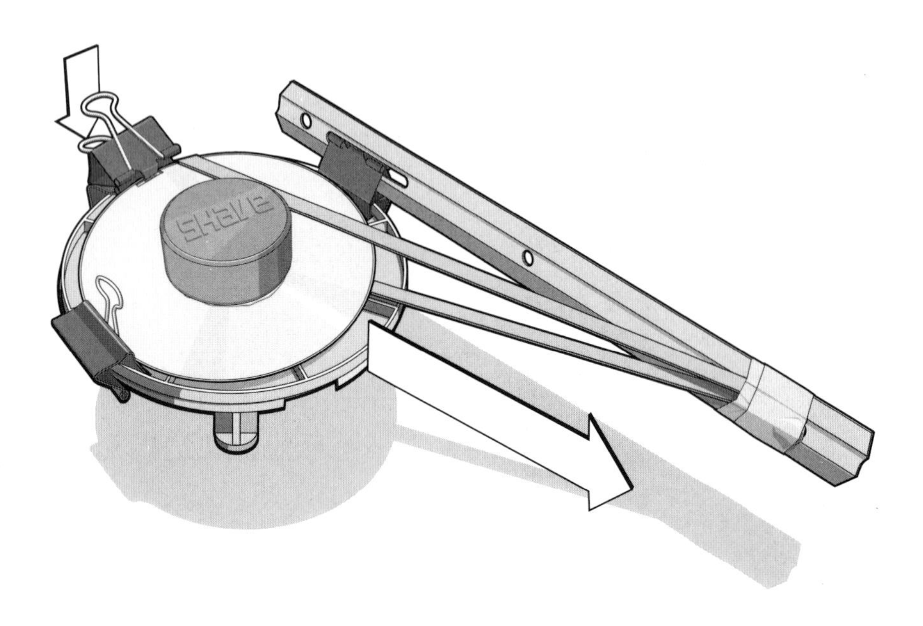

Before loading the Bowler Hat Launcher, put on your safety glasses, just in case!

Clip the disc onto the top binder clip with the modified tooth detail. This clip should be fully shut, ensuring the disc will stay locked into place until you release it. Adjust the disc custom notch-out if it's too small.

Once the disc is locked in, stretch the rubber band around to the near side until it stays put behind the disc, as shown. Having the rubber band offset will create a perfect Frisbee rotation when released.

Aim and push down on the metal clip handle to launch. Remember, this is a homemade launcher—***be prepared for the unexpected.*** If the ruler arm is not secure, it could whip back and hit you. Be careful! The disc launcher is designed to launch the disc forward in an accurate direction; however, you may need to adjust the design for best results. Always wear safety glasses and never shoot at a living target.

PAPER THROWING STAR

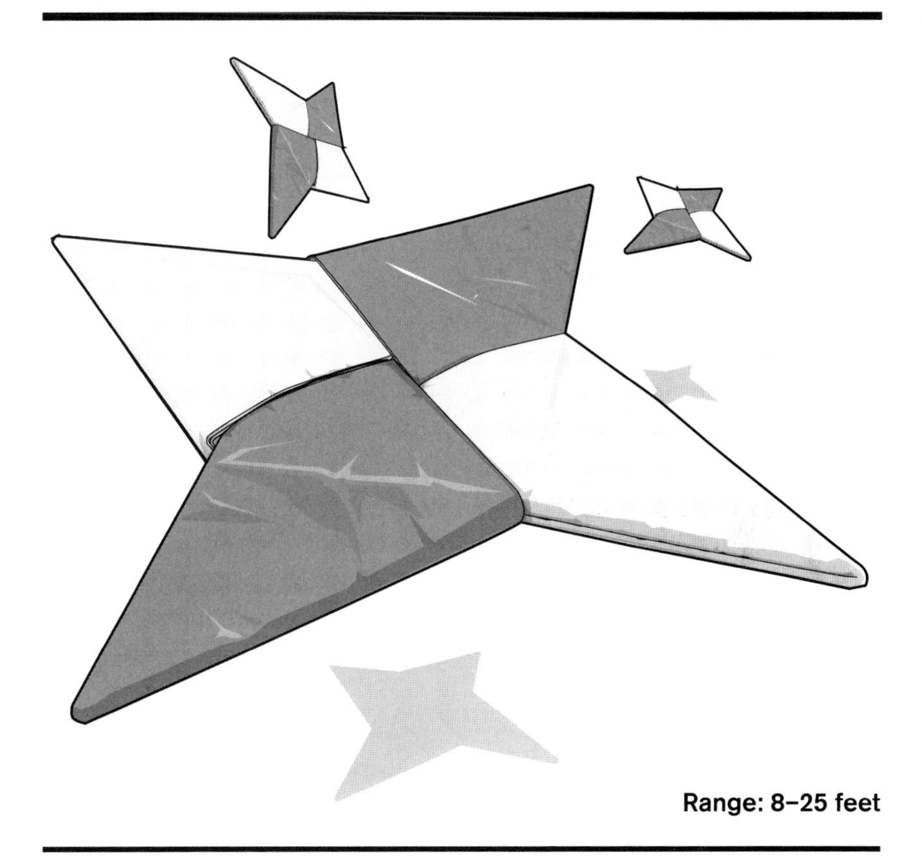

Range: 8–25 feet

Wielded by the ninja as a secondary weapon, the Paper Throwing Star (also known as a Shuriken or Ninja Star) can play a pivotal role in a secret agent battle. Using the ancient art of origami, a skilled warrior can manufacture various sizes and colors. It's a simple weapon, but you will need some patience, little grasshopper. With no cost, this will be a perfect addition to your MiniWeapons arsenal.

Supplies

2 sheets of paper

Tools

Scissors
Pencil

Step 1

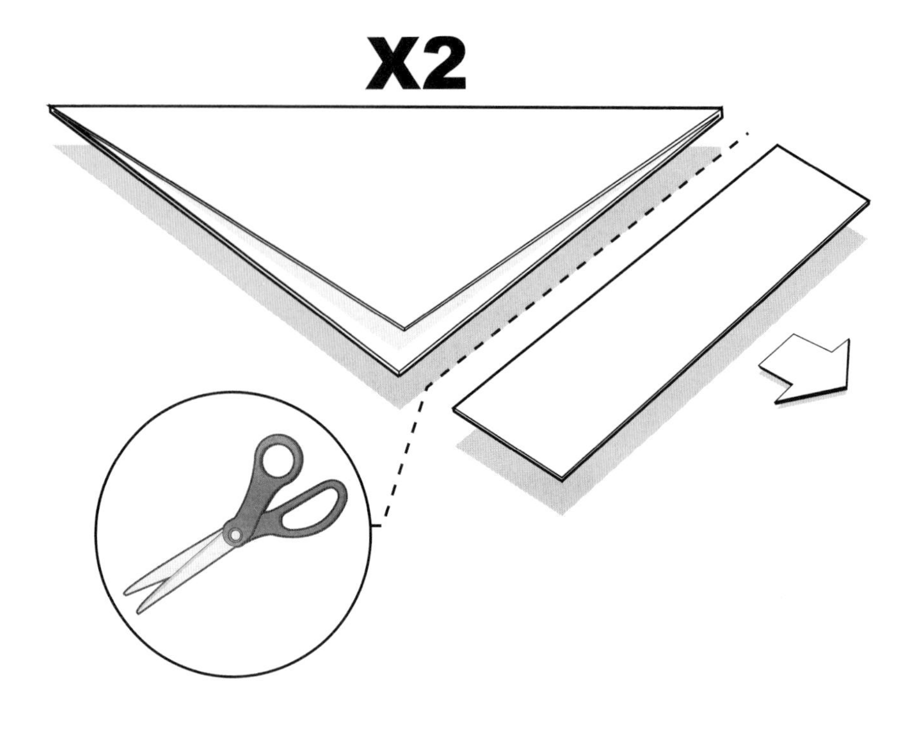

This ninja star design is crafted from two identical, perfectly square sheets of paper.

If you are starting with two standard sheets of rectangular printer paper, fold over one corner flush to the opposite side and then crease the diagonal fold. This is sometimes called a valley fold. While the paper is still folded, use scissors to cut along the layered edge to remove the leftover rectangle strip.

Repeat this step with one more sheet of paper so you end up with two identical squares.

Step 2

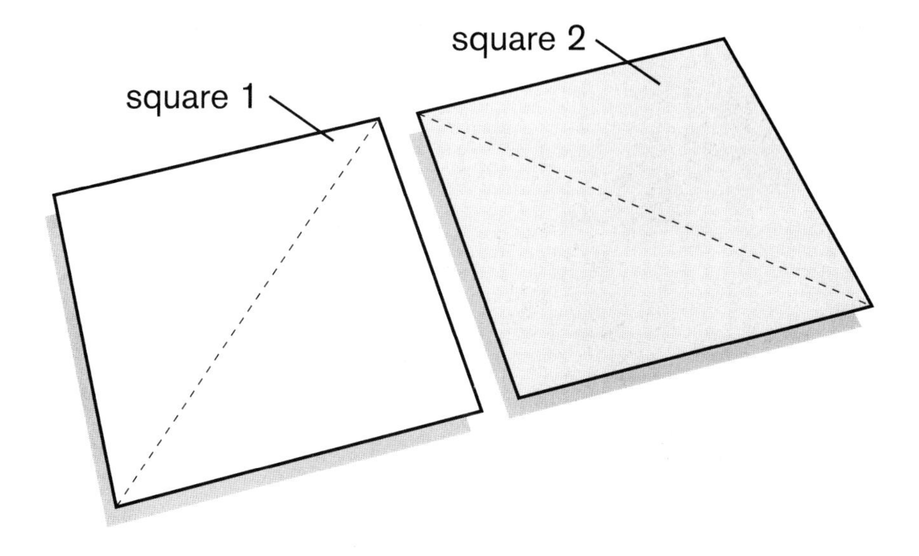

Unfold both sheets of paper to reveal the custom squares. Mirror the sheets next to one another as shown in the illustration. Make sure the folds are in the correct direction.

Step 3

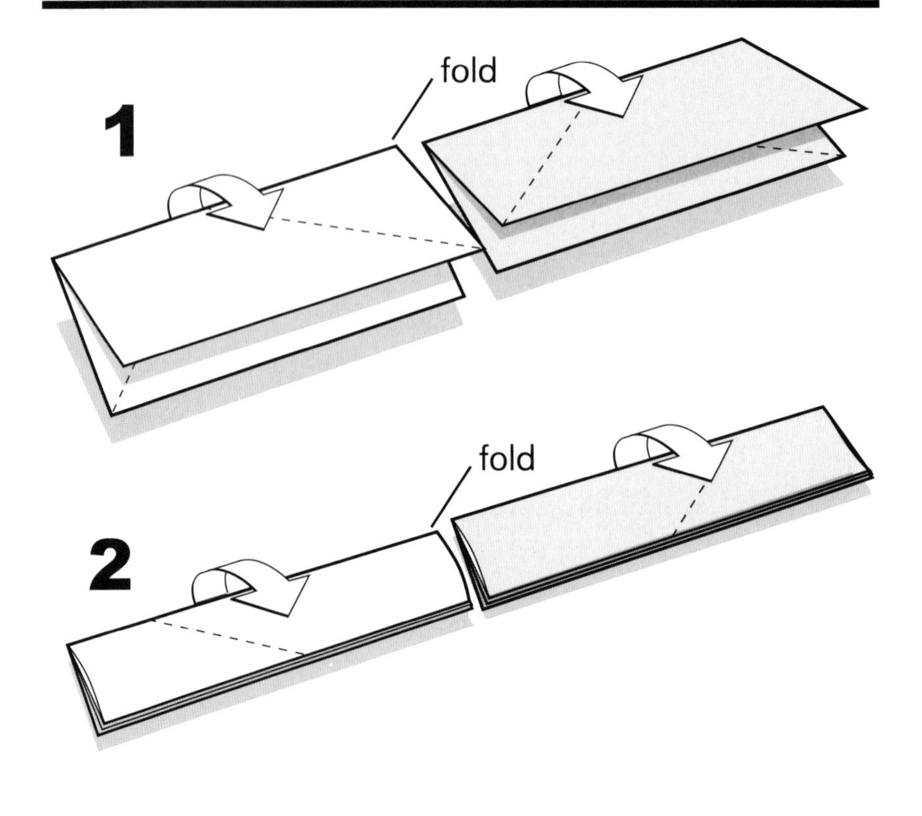

Fold both squares in half to create two rectangles (1). Press your finger along the crease lines make them flat and sharp. Then fold each sheet once more, making two long rectangles (2). Each rectangle will be four layers thick.

Step 4

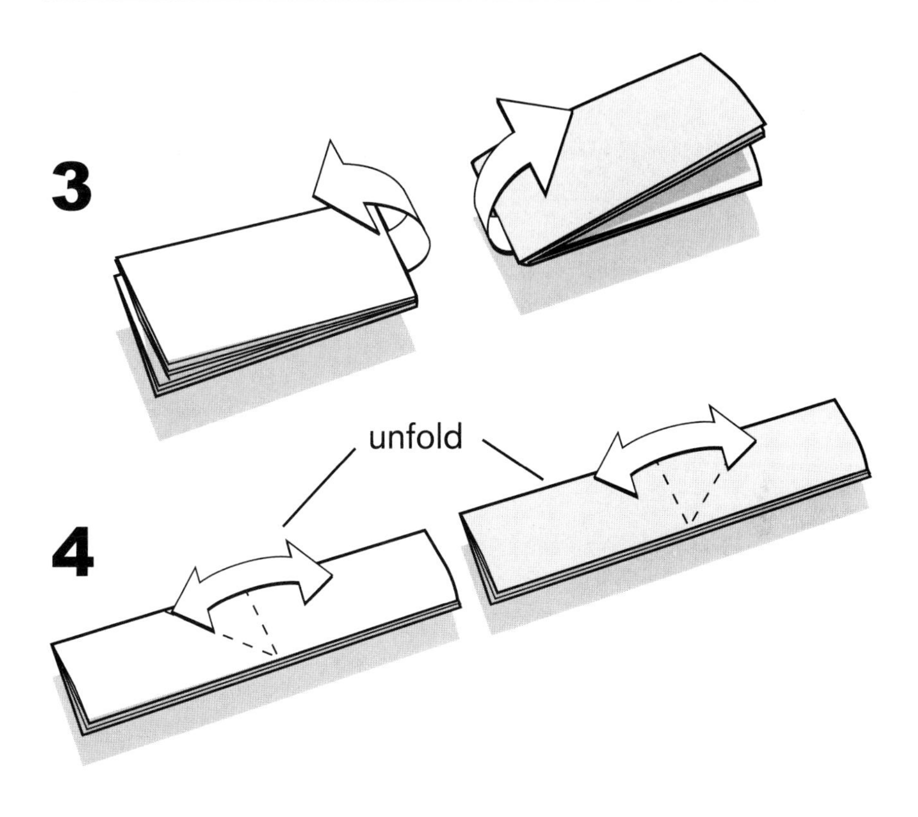

Next, fold both rectangles in half, end to end, as shown (3). These folds will create crease lines that will help with the next step. After you have folded both rectangles, unfold them to reveal the crease lines (4). You may want to use a pencil to highlight the new crease lines.

Step 5

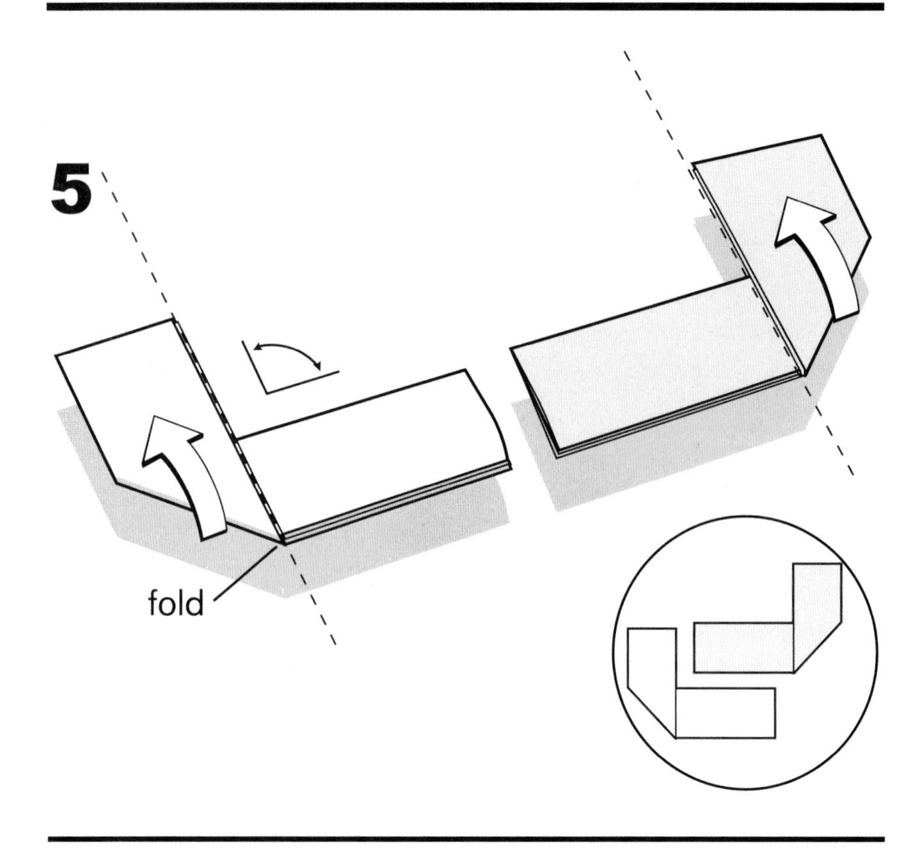

Using each middle crease line, valley fold one end of each rectangle at an overlapping 45-degree angle (5). Apply pressure to the fold lines to hold the "L" shapes in place. Note that the second folded "L" should mirror the first. Compare your results with the circled diagram and adjust if needed.

Step 6

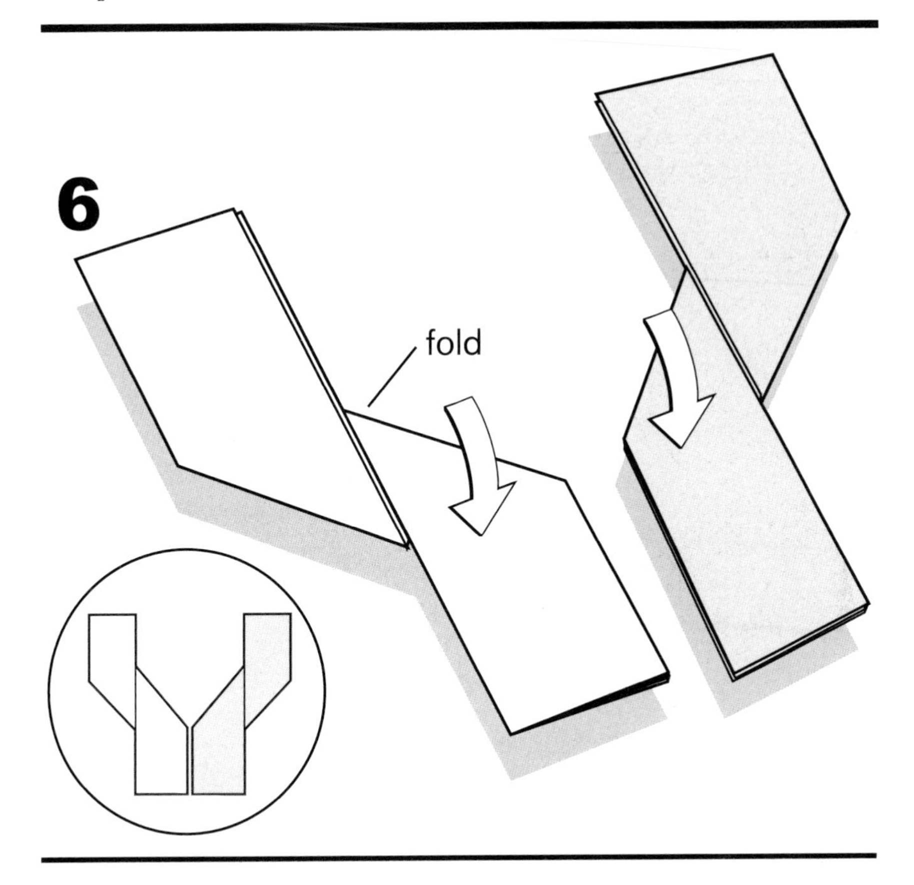

Now mirror step 5, and repeat the valley fold on both pieces of paper. Fold down the "L" at a 45-degree angle so that the new fold runs along the first fold at the center (6).

Repeat this step on the second "L." Compare your results with the circled diagram and adjust if needed.

Step 7

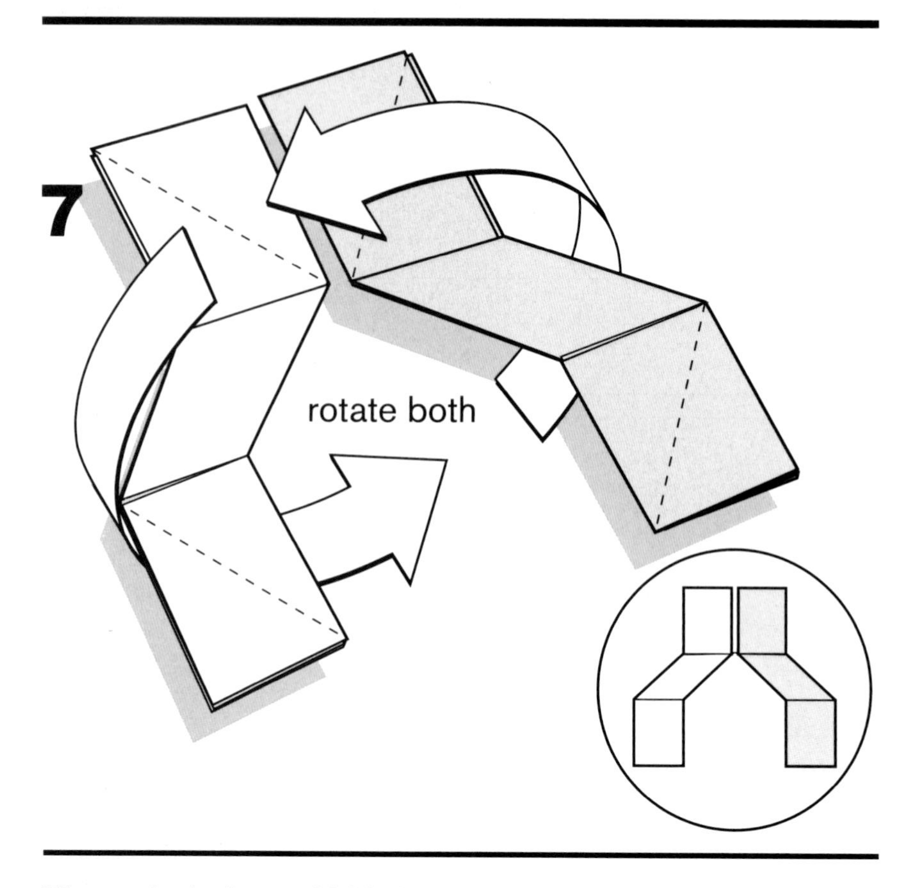

Flip over both pieces of folded paper and line them up as shown in the diagram.

The dotted lines in this illustration represent the next step's fold lines. If you want to mark those lines with a pencil to make them easier to follow, do that now.

Step 8

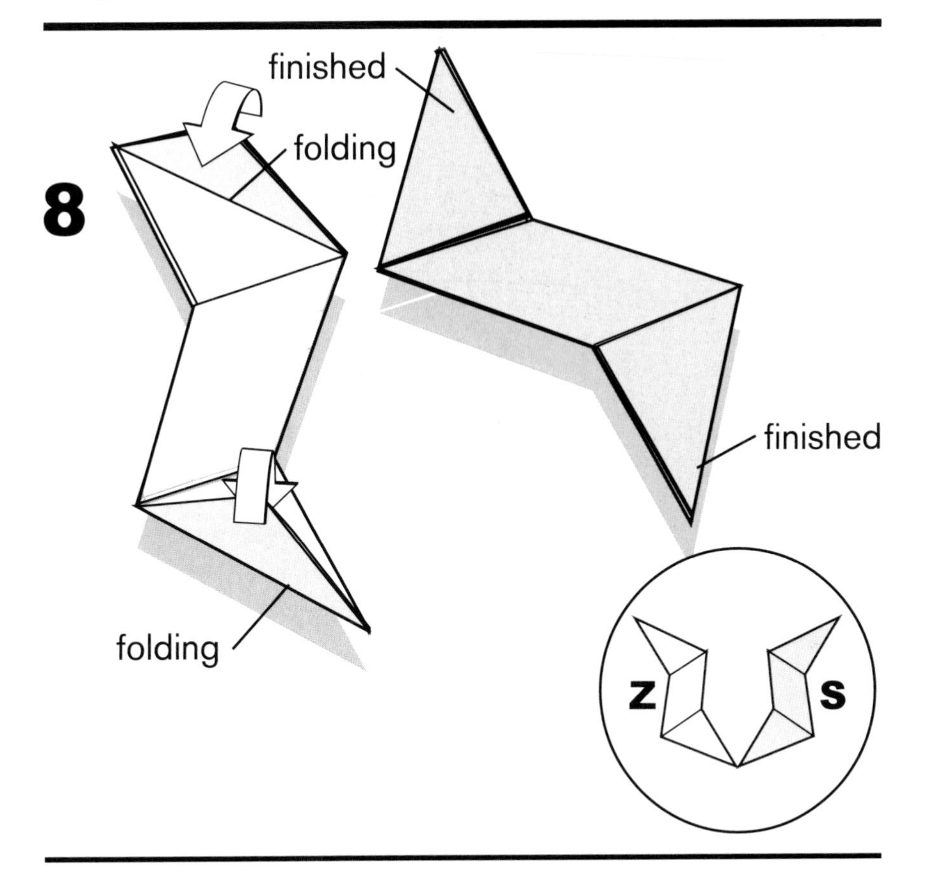

In this step, both pieces of paper will be folded until they resemble a "Z" and an "S." You will accomplish that by working with the square segments and the ends of each paper assembly.

On the first folded piece of paper, valley fold the end squares to create two triangles. Use the circle diagram on the bottom right to compare your folds. Once they are correct, use the diagram to line up the "Z" and "S" pieces.

Step 9

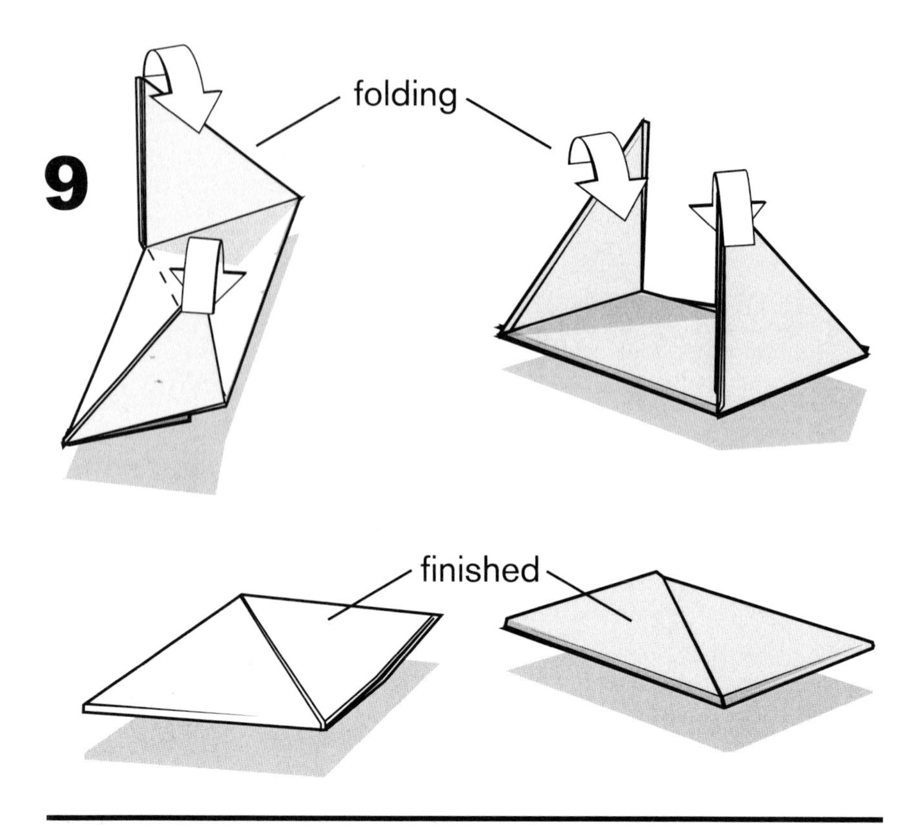

In this step, you will create crease lines to be used later in step 11.

Take the folded tip triangles and fold them toward the center to create a parallelogram (bottom image).

Repeat this step on the second fold assembly. Once completed, *unfold* both pieces of paper so they resemble a "Z" and an "S" again, just as in step 8.

Step 10

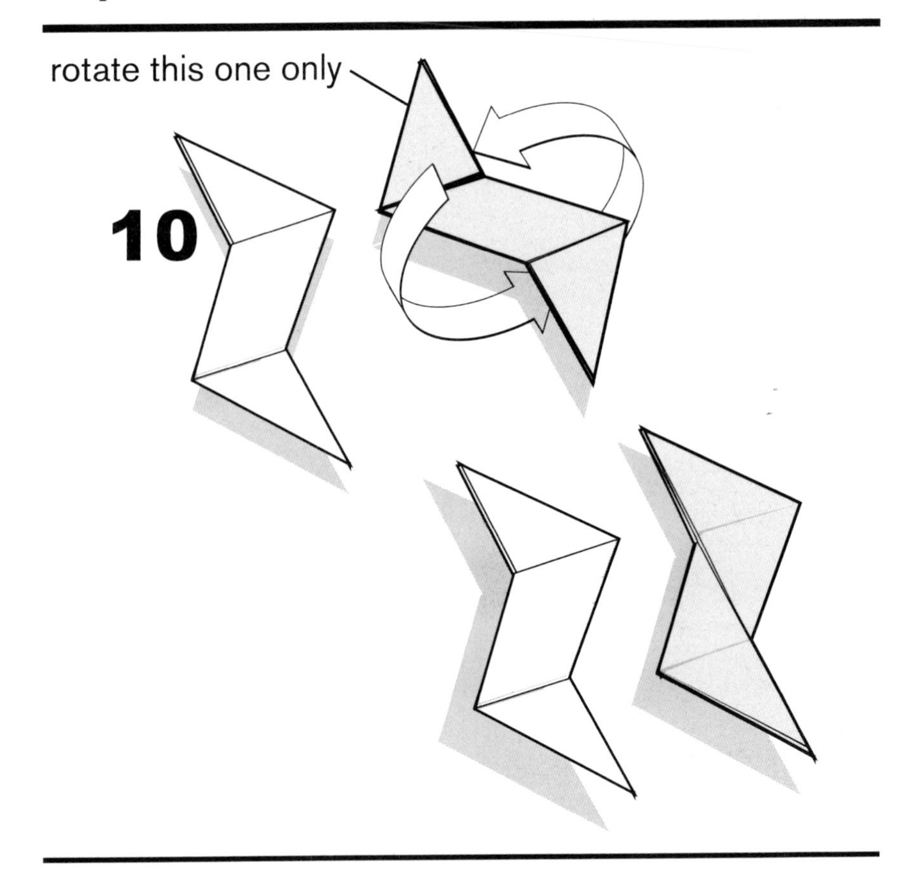

Flip only the "Z" piece of paper (on the right in the top image). Line up both halves of the ninja star similar to the bottom illustration. Though the outer shapes are similar, the inner folds now aren't.

Step 11

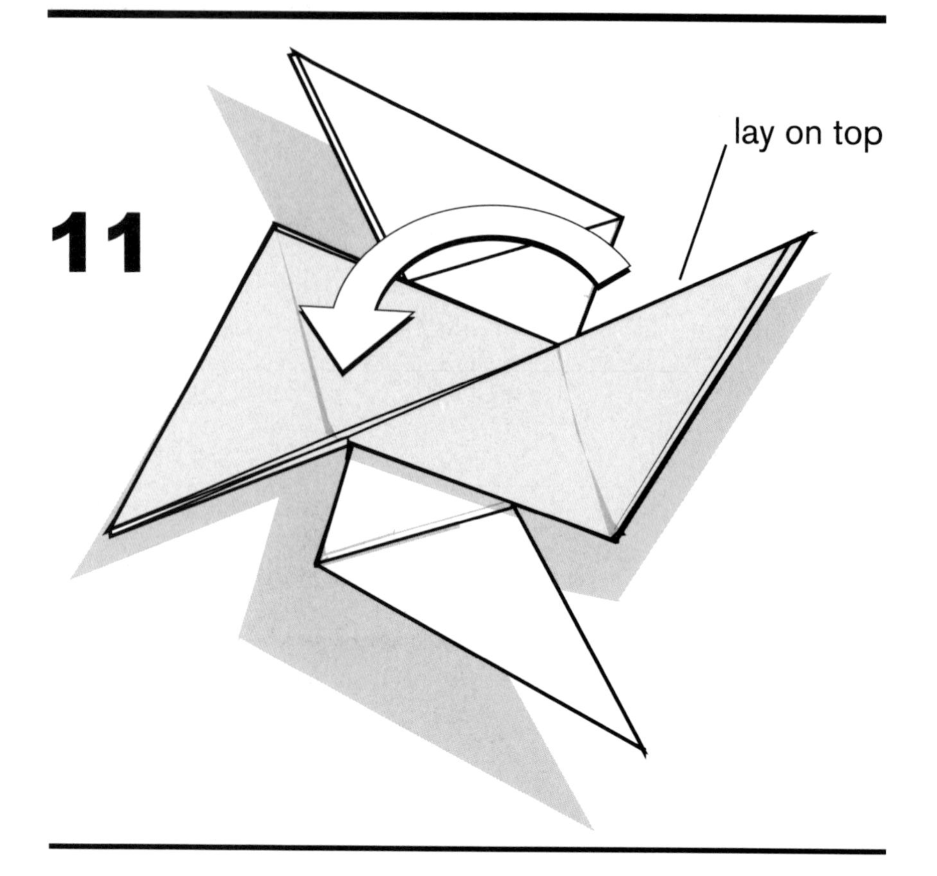

Rotate the second (right) paper assembly 90 degrees and place it on top of the first paper assembly, as shown.

Step 12

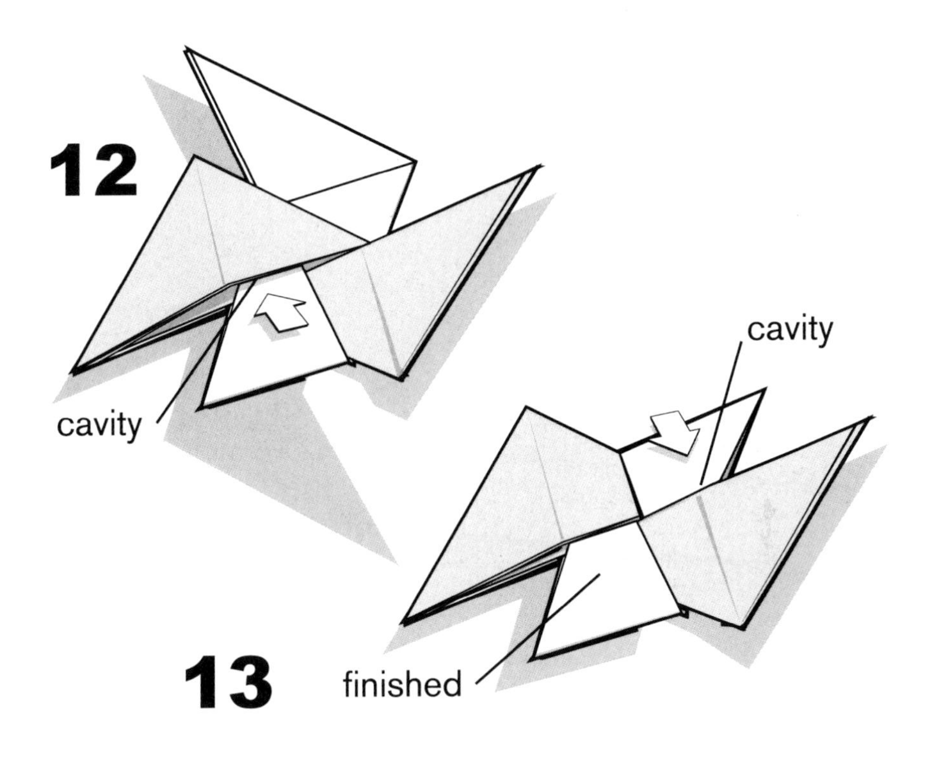

Fold the top and bottom points of the bottom paper assembly into the cavities of the top paper assembly. See the illustration.

Step 13

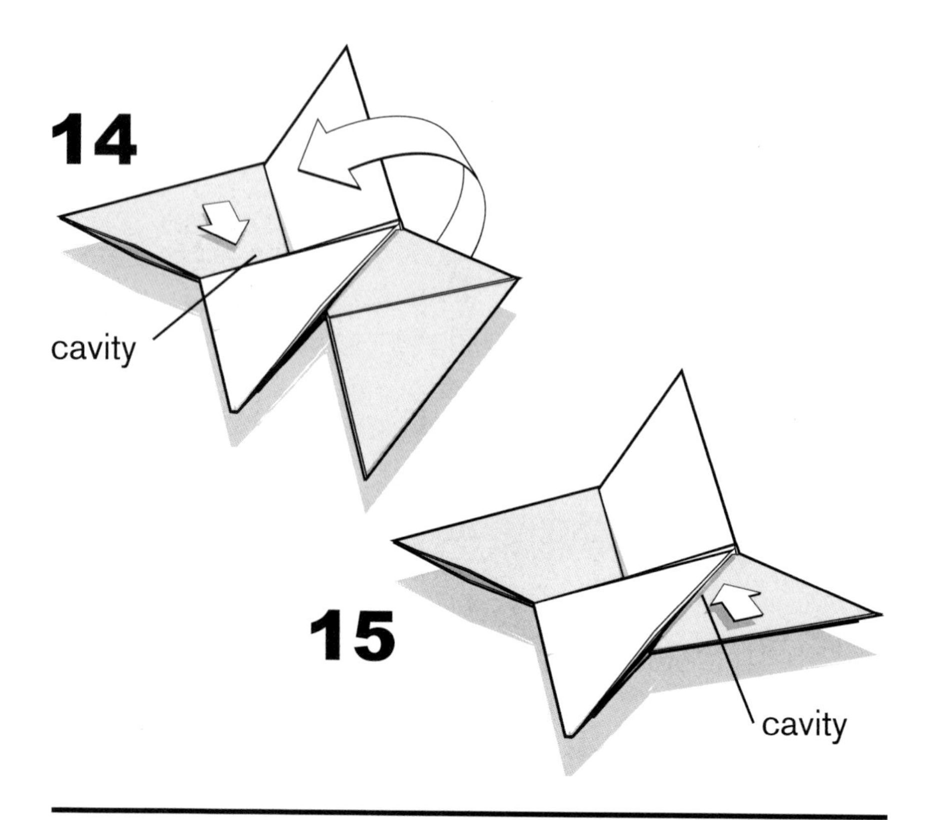

Now, flip over the entire, partially assembled Paper Throwing Star and complete the folds on the backside. Once flipped, the second paper assembly will be on the bottom. Fold the top and bottom points into the cavities of the top paper assembly as you did in the previous step. Make sure all the points are tucked in and secure. Your Paper Throwing Star is complete! To toss, grasp it with two fingers and fling it like a Frisbee. ***Remember: never throw your paper star at a living target. Its paper points could cause eye damage.***

4 GADGETS

PAPER DART WATCH

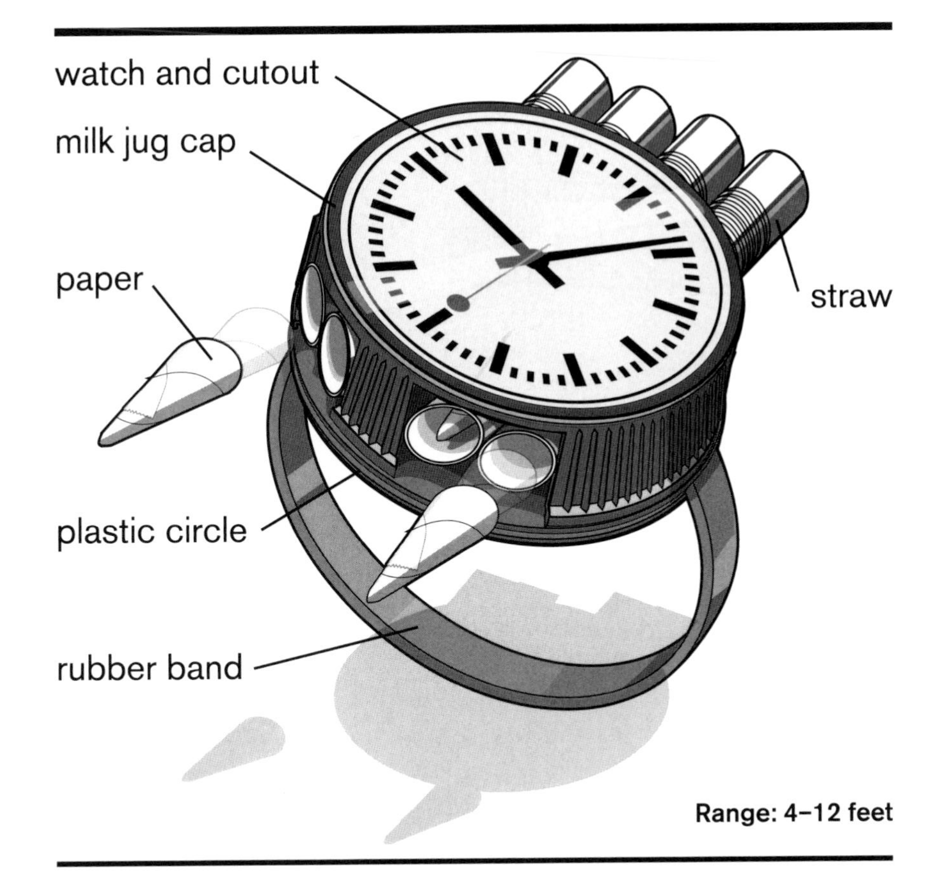

Range: 4–12 feet

The Paper Dart Watch is standard issue for any agent involved in covert intelligence. Its integrated elastic band keeps it locked to your wrist when you're chasing somebody . . . or being chased! Lock onto your target by rotating the watch face, then launch one of the four paper darts. You'll soon understand why this gadget is an agent's best friend.

Supplies

1 sheet of 4-inch-by-4-inch plastic scrap
1 plastic milk jug cap
4 drinking straws with flexible heads
1 1-inch brass fastener
1 large rubber band
1+ lifestyle magazines

Tools

Safety glasses
Marker
Scissors
Hobby knife
Hot glue gun

Ammo

1+ sticky notes
Clear tape

Step 1

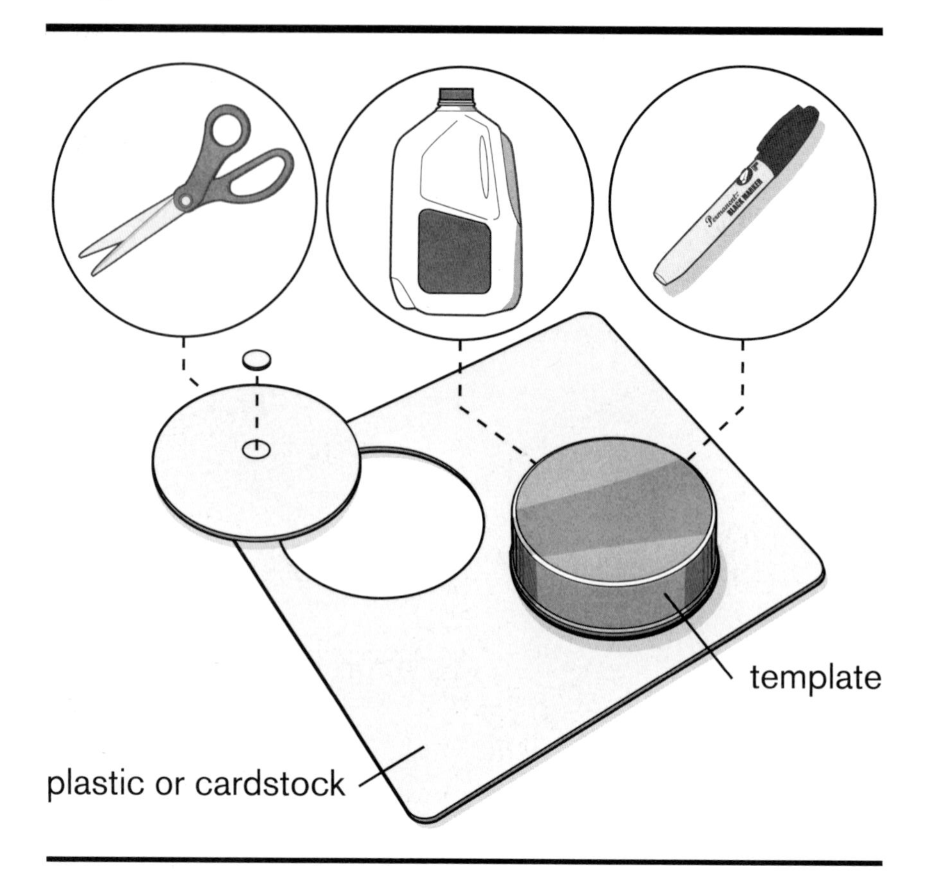

First, locate some flat, thin but rigid plastic material, such as that used for product packages with clear windows. If plastic is unavailable, heavy-duty cardstock will also work. You will need two 2-inch-by-2-inch squares, or one 4-inch-by-4-inch square, of whatever material you decide to use.

Place a milk jug cap on top of the thin plastic, then trace the cap's diameter onto the plastic with a marker. Do this twice to make two circles. Use scissors to cut out both plastic circles.

In one of the circles, poke a small opening through the center of the disc. The diameter of this circle will need to fit a brass fastener, so smaller is better.

Step 2

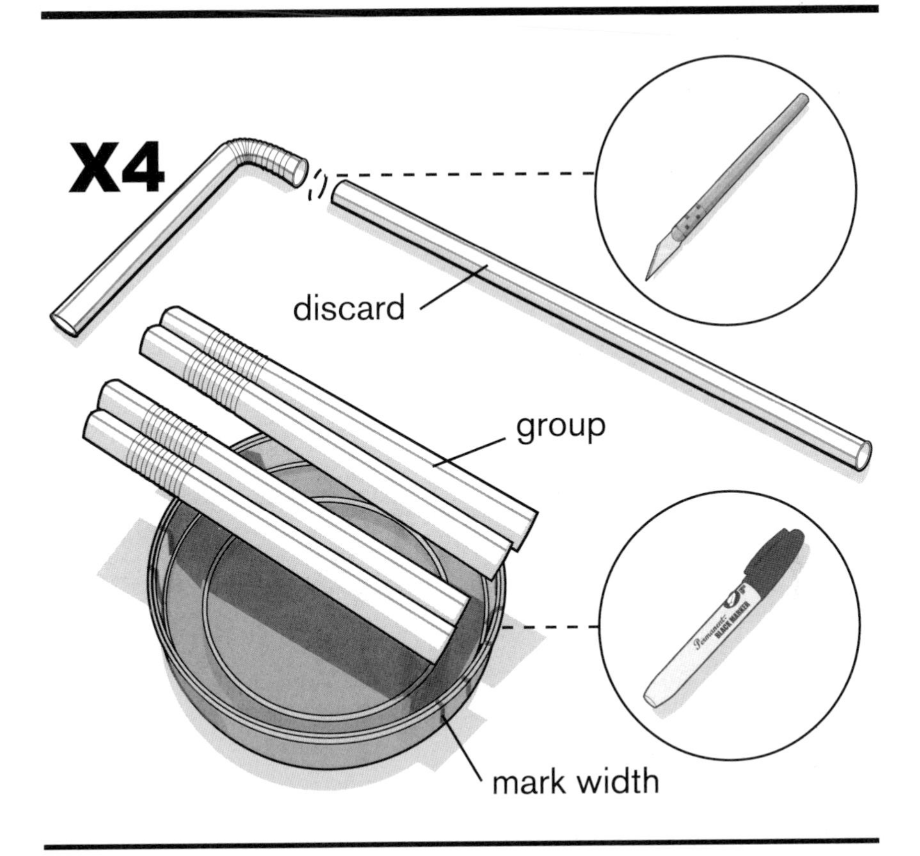

Using a hobby knife or scissors, cut the bendable heads off of four plastic drinking straws, just below the bellowed (bendy) sections. Discard the longer segments of the straws.

These short straws will eventually be placed inside the milk jug cap in step 3. But first you will first mark the area of the milk jug cap that you will need to remove to fit the straws inside. Place two pairs of straws next to one another on the top of the underside of the milk jug cap, as shown. Use a marker on the milk jug cap to mark the area you will cut out. Do *not* place the straws in the center of the cap because they will interfere with the brass fastener rotating feature.

Step 3

Using a hobby knife, cut along the eight lines you marked on the side of the milk jug cap, then remove the tabs as shown in the upper right diagram. In addition, cut a small hole in the center of the cap. This hole will eventually need to fit the brass fastener. Remove any other plastic that might interfere with the straws fitting inside the cap.

Now place all four straw sections into the milk jug cap and secure them in place using hot glue, as shown. The bellowed sections should overhang one side of the cap. These flexible tubes will make firing the darts easier.

Step 4

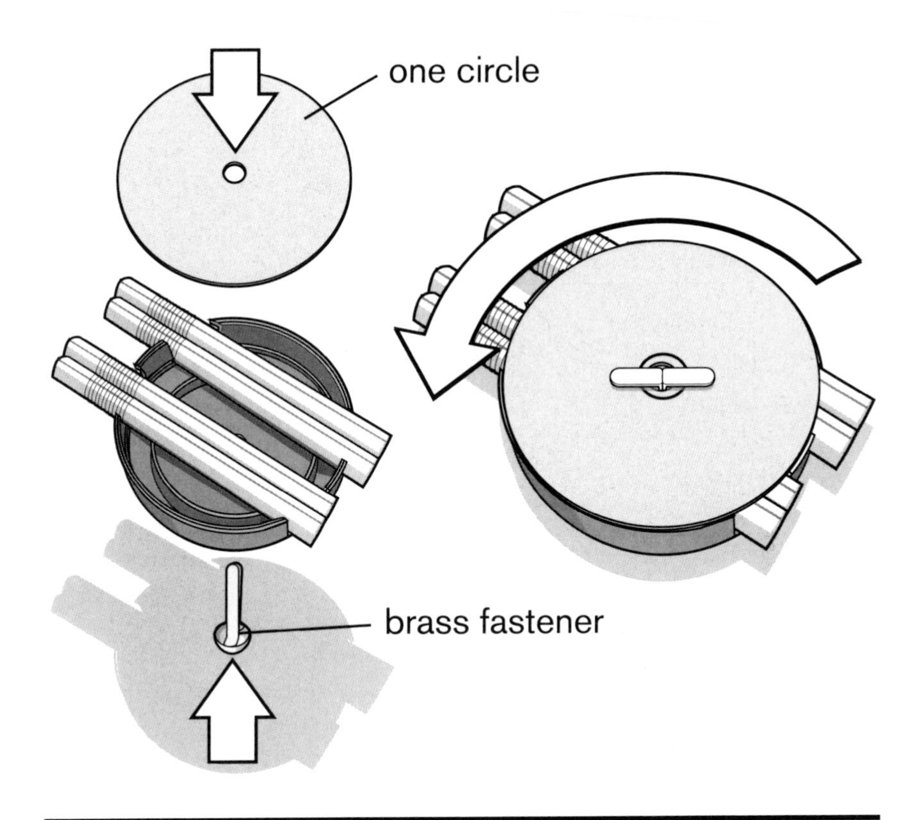

From the outside of the milk jug cap, slide a brass fastener through the center hole in the cap assembly and up through the plastic circle with the center hole. Bend the brass fastener into place.

Do not hot glue the plastic circle to the straws or milk jug cap; this will prohibit the Paper Dart Watch from rotating.

Step 5

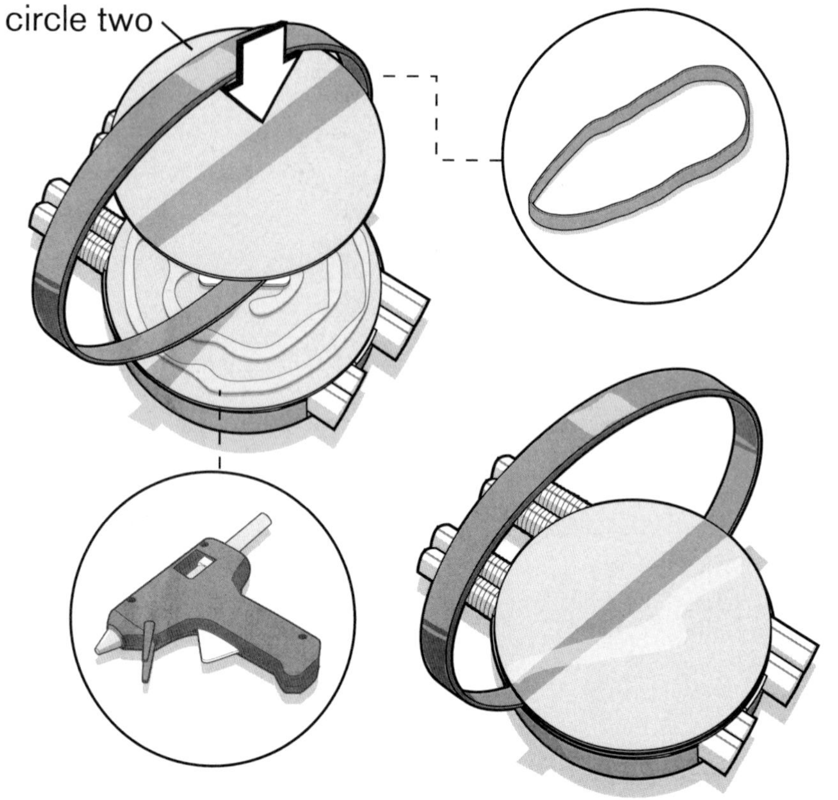

Next, test a wide rubber band around your wrist to see if it fits comfortably, and then remove it. If the band's fit is satisfactory, sandwich the rubber band between the two plastic circles using hot glue.

Step 6

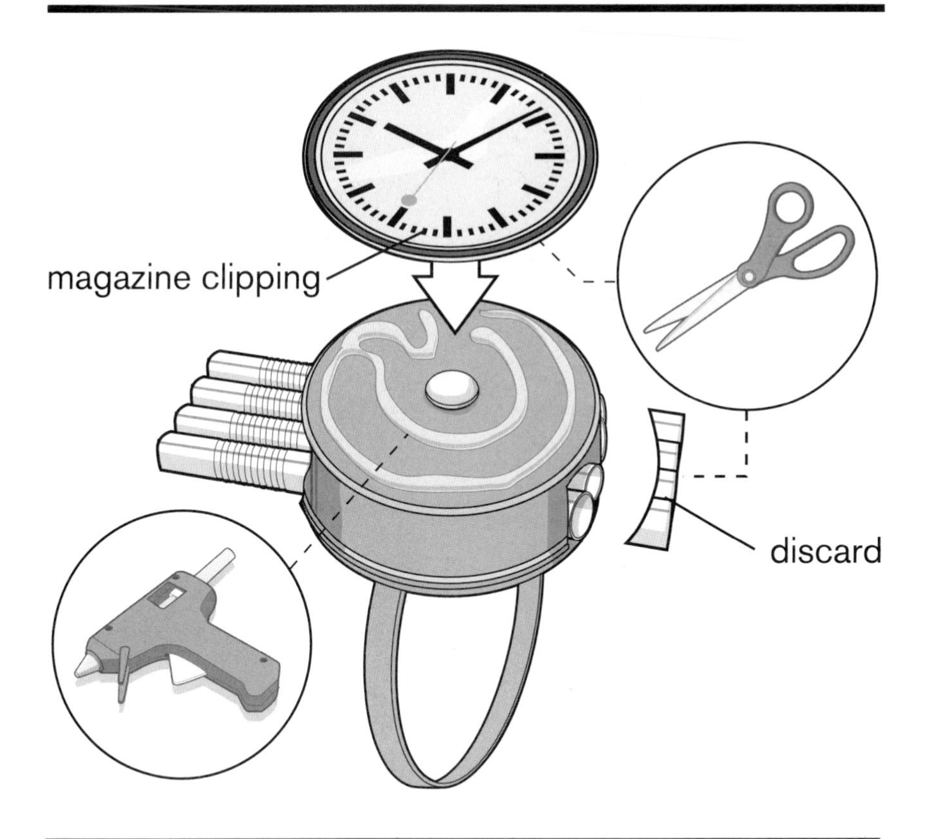

You will need a watch face to complete your gadget's disguise. Flip through a few lifestyle magazines for watch ads you can cut out. Many of these ads will have life-size or larger-than-life-size, high-resolution pictures of watches. Find a picture of a watch face that fits the milk cap diameter perfectly. You can also draw a watch face or print one from an online watch ad.

Hot glue the watch face to the top of the milk jug cap assembly, as shown. This realistic face will help camouflage your gadget from curious eyes.

Finally, use scissors to trim down the straws, but only on the non-bellowed ends. The flexible straw ends should remain on the watch so they can be positioned for optimal firing.

Step 7

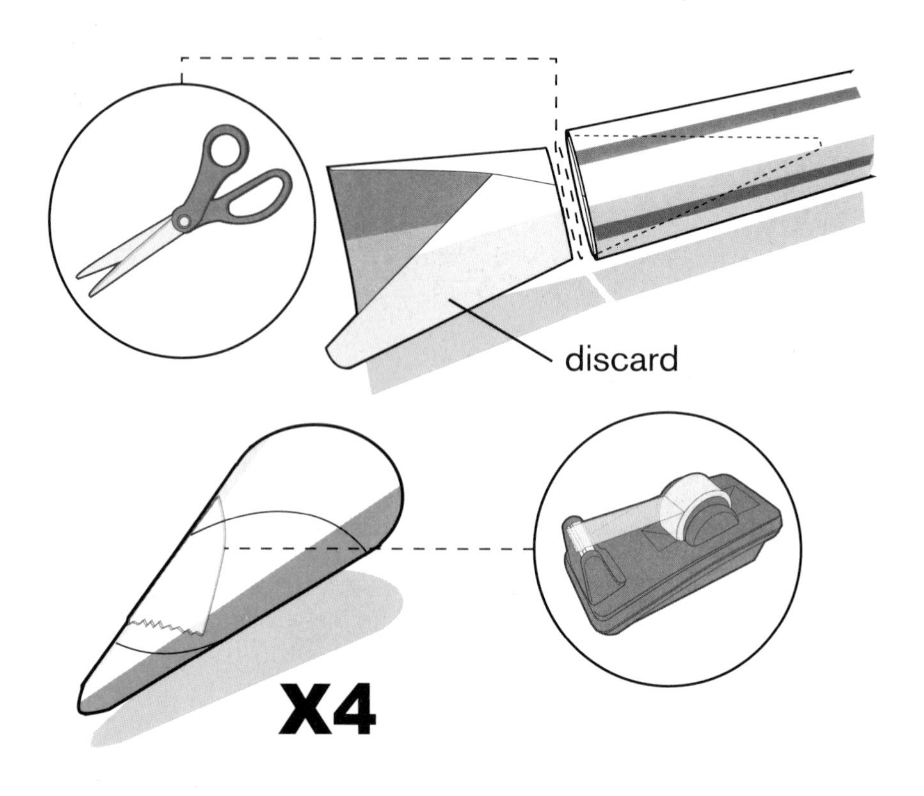

The four darts for this MiniWeapon are constructed from paper. Start with 1½-inch-square pieces of paper. If you have a 3-inch-by-3-inch sticky-note, fold it in half, then fold it one more time to create a square. Then unfold it and cut along the crease lines to create four 1½-inch-square pieces (not shown).

Roll each small paper square into a cone. Once you have a cone shape, place a small piece of clear tape on the edge to hold it in place.

Insert the small paper cone into one of the straws, being careful not to damage it. Trim off any extra paper with scissors. Now the dart's maximum width is the same diameter as the straw. Repeat to make three more darts.

Fit one paper dart into each of the straws, with the tip facing the exit (nonbellowed end). Adjust a straw neck, take a deep breath, and exhale sharply into the straw to fire the dart. Always be responsible and fire the darts in a controlled manner. ***Safety glasses are a must, as is staying clear of spectators.***

GRAPPLING GUN

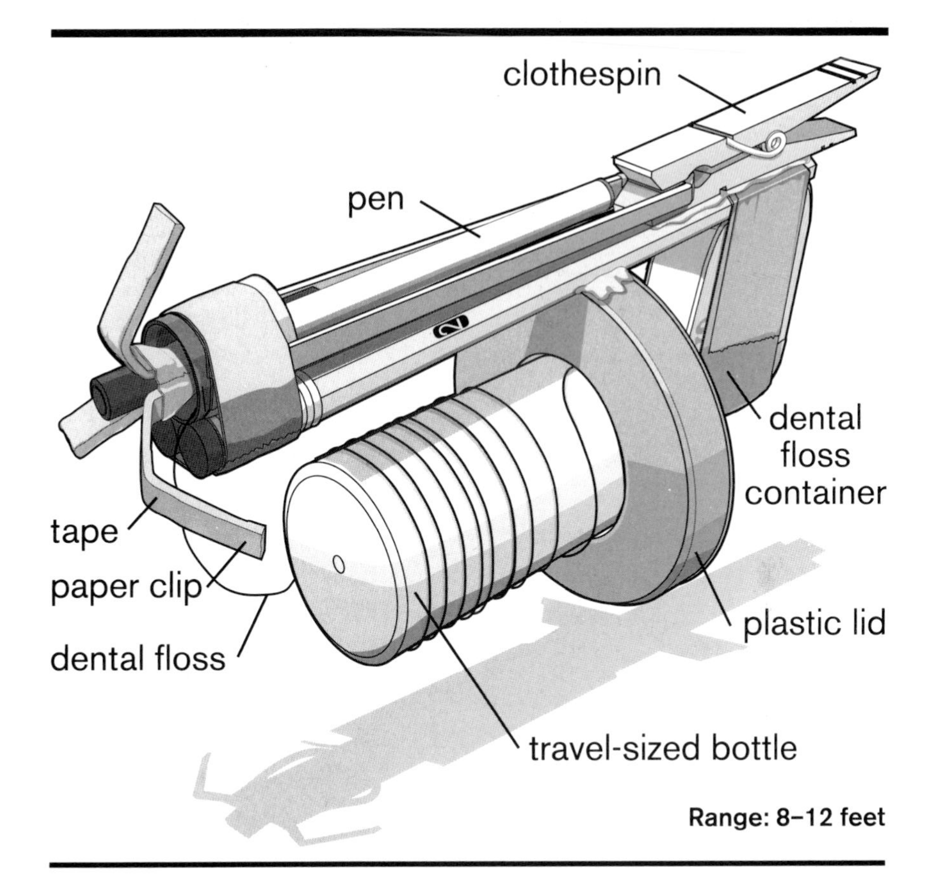

Range: 8–12 feet

The tactical Grappling Gun allows you to snag something beyond your grasp or swing over obstacles such as trip-wire-fused land mines. It can be a tricky project, but with a little patience and some adjustment you'll soon be shooting, grabbing, and snatching whatever you please.

Supplies

3 large paper clips
Duct tape
1 plastic ballpoint pen
1 standard art marker
1 wide rubber band
2 pencils
1 plastic package of dental floss
1 clothespin
1 travel-size plastic bottle (approx. 3.4 fl. oz. or smaller)
1 3-inch soft plastic lid

Tools

Safety glasses
Pliers
Hobby knife
Hot glue gun

Step 1

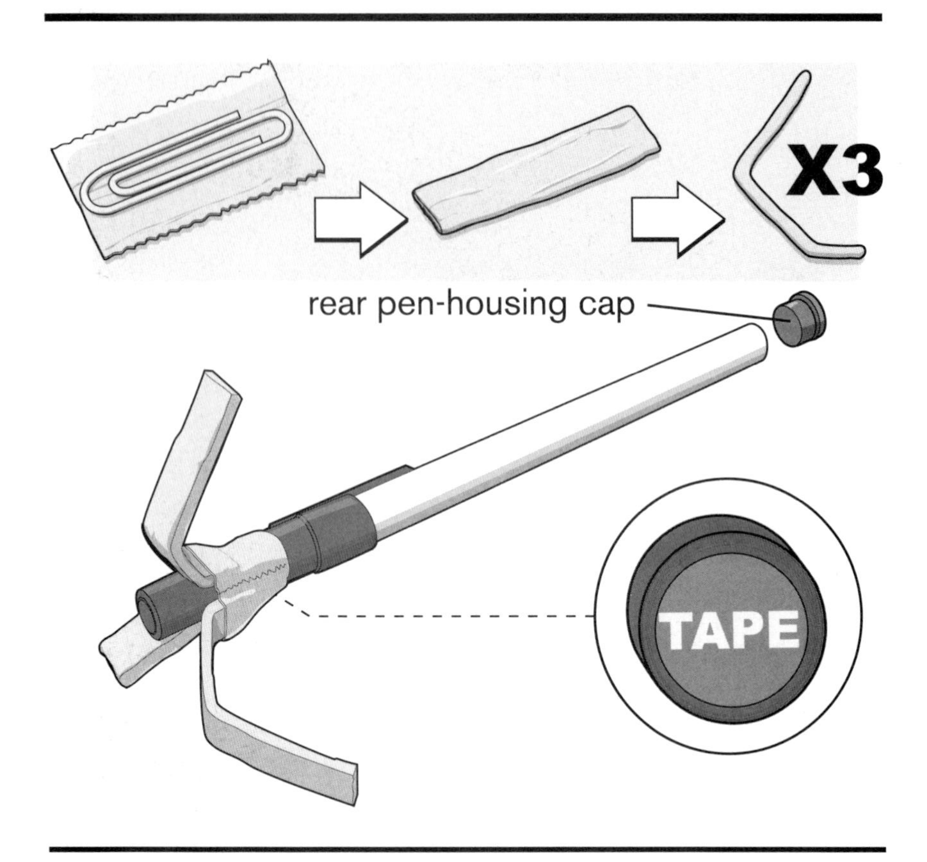

The gun's grappling hooks are constructed from three large paper clips. Each hook is made separately and then attached to the pen housing.

To construct one hook, place a large paper clip on a slightly larger piece of duct tape, then securely wrap the clip until it is completely covered with tape. Then, use pliers to bend the covered paper clip in two spots. Refer to the illustration for the position of the bends. Repeat this step with two more paper clips.

Evenly space the hooks around the cap end of a ballpoint pen, securing them with duct tape. Finally, use pliers to remove the rear pen-housing cap. *Do not* discard this cap; you will be reattaching it in a later step.

Step 2

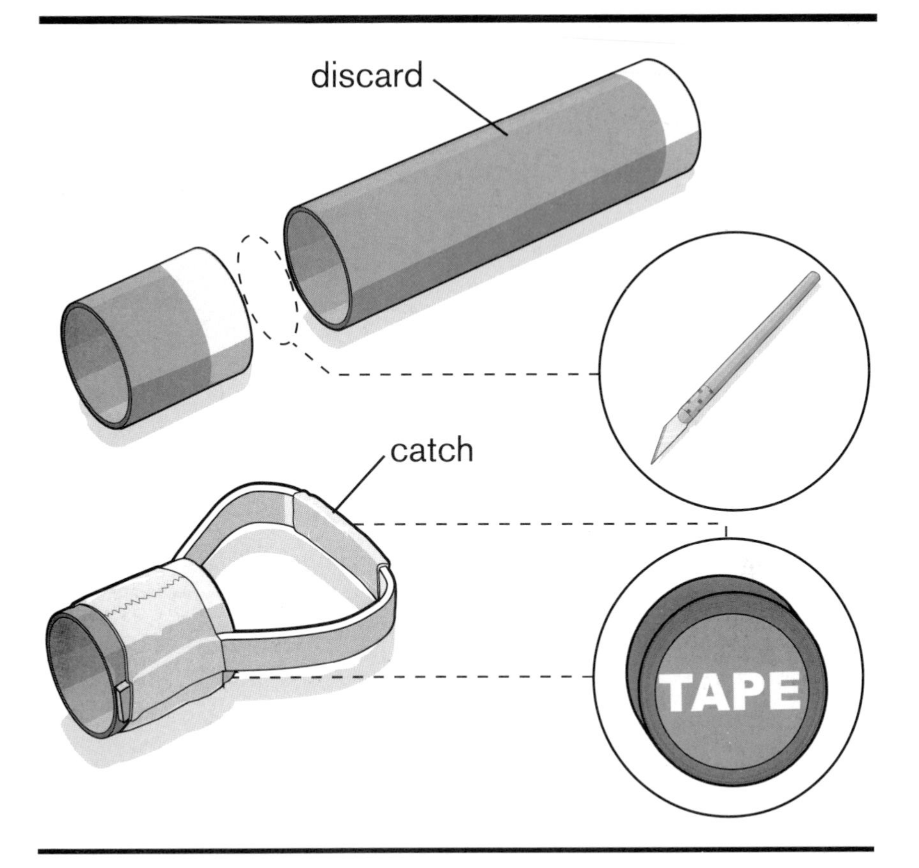

The grappling hook guide—the barrel—is fabricated from a standard art marker or similar diameter container (such as a lip balm cap) that is slightly larger than the pen grappling hook. If you use a marker, one made from recycled plastic is best, because the material is relatively soft. However, any marker will work. Using pliers, remove the marker's end cap by slightly rotating it from the housing. Remove the marker nib using pliers, then slide it out the tip of the marker. Discard this nib to avoid a mess.

With a hobby knife, pierce the housing to remove a 1-inch cylinder; rotating the housing while the knife is piercing the plastic will make it easier. Discard the internal ink cartridge. (If the housing contains ink residue, rinse it out and dry it.)

Duct tape a wide rubber band to the 1-inch cylinder. Then wrap an additional piece of duct tape around the back of the rubber band to create a finger catch.

Step 3

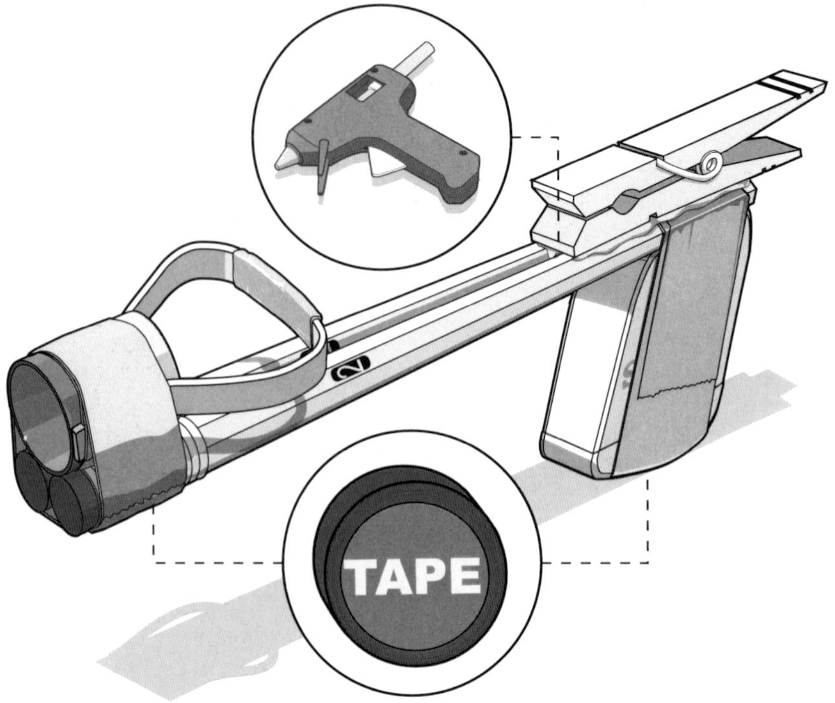

Start building the gun's frame by duct taping two pencils together. Then, on the eraser end, tape the Grappling Gun barrel with its attached rubber band.

On the opposite end, attach the gun's handle, crafted from an empty dental floss container. First, pop out the insides of the floss container but do not discard the floss; you'll need it. Duct tape the container onto the two-pencil barrel as shown. (If you do not have a dental floss container, experiment with other health and beauty packages.)

Finally, hot glue one clothespin on top of the pencil barrel above the dental floss handle.

Step 4

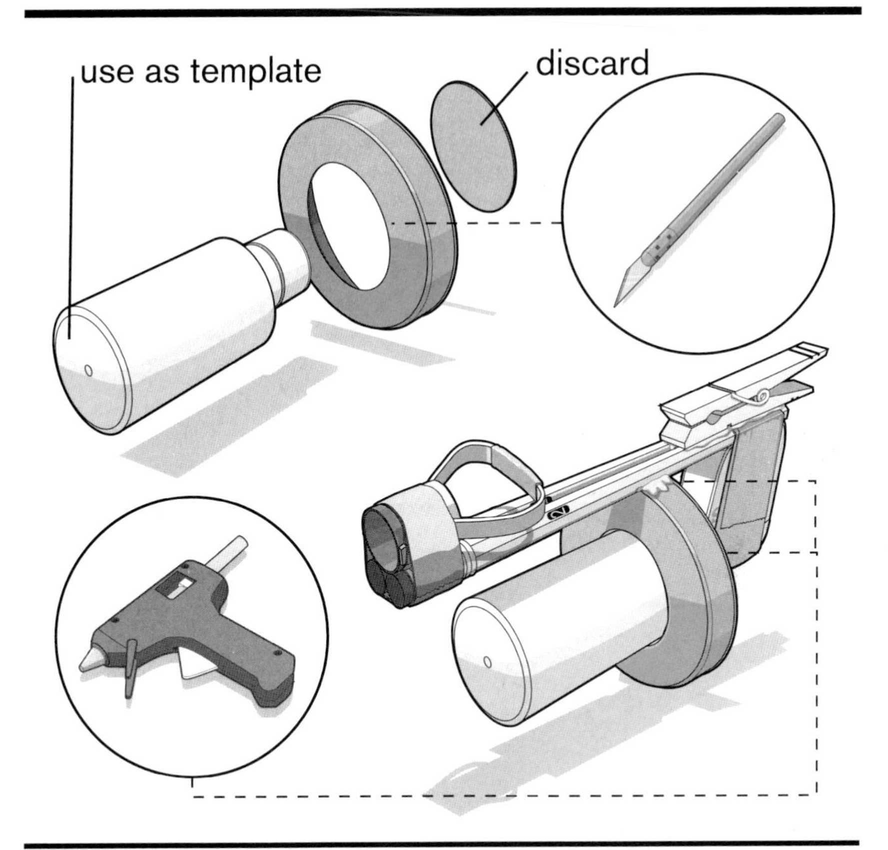

Next, assemble the lower-mounted Grappling Gun rope holder. It's built from a small, travel-size plastic bottle and a soft plastic lid approximately 3 inches in diameter (the top from a can of Pringles works well). If this lid size is unavailable, just cut a 3-inch-diameter circle from cardboard stock or experiment with other sized lids—for example, coffee can, cream cheese, or peanut butter lids.

Using the small bottle as a template, trace and cut its diameter from the center of the lid. Then slide the small plastic bottle into the opening and hot glue it into place.

Affix the lid-and-bottle assembly to the dental floss handle and double pencil frame carefully using hot glue, as shown.

Step 5

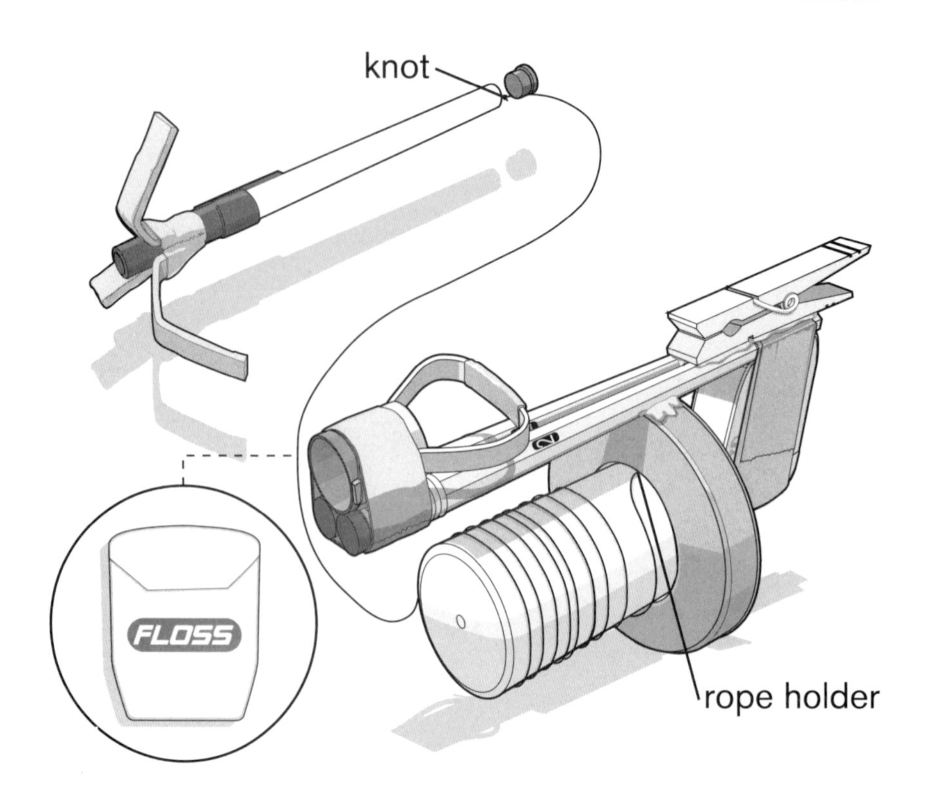

Uncoil about 10 feet of dental floss (or thread). Tie a small knot to one end and insert the knot into the back of the pen housing, then seal it with the original rear pen-housing cap, securely sandwiching the knot inside the housing.

Finally, tie, tape, or hot glue the other end of the floss line to the back of the Grappling Gun rope holder. Loosely coil the remainder of the floss around the bottle, working outward.

To fire, snap the rubber band into the clothespin, then slide the grapping hook and its attached line into the marker barrel as far as it will go. Aim the Grappling Gun and press the clothespin trigger when you're ready.

Be aware that this is homemade projectile launcher is capable of malfunctions and misfires. Adjust it if needed, and always fire away from living targets.

ASSASSIN BLADE

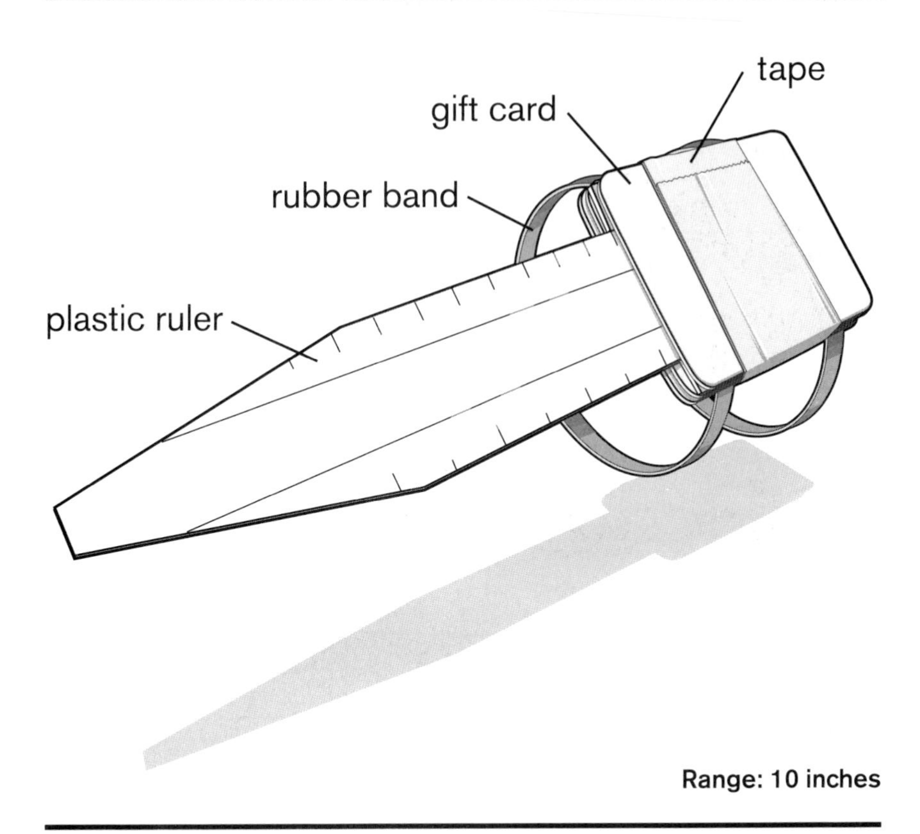

Range: 10 inches

The Assassin Blade is a nonlethal, retractable wrist weapon that allows you to keep a low profile when pursuing your target up close. It's designed to be a quick-strike MiniWeapon that stays out of view. Then, with a flick of your wrist, the ruler blade springs out of your sleeve and is ready for the attack.

Supplies

3 expired or zero-balance plastic gift cards
1 plastic ruler
1 brass fastener
1 small binder clip (19 mm)
Duct tape
3 large rubber bands

Tools

Scissors or hobby knife
Hot glue gun

Step 1

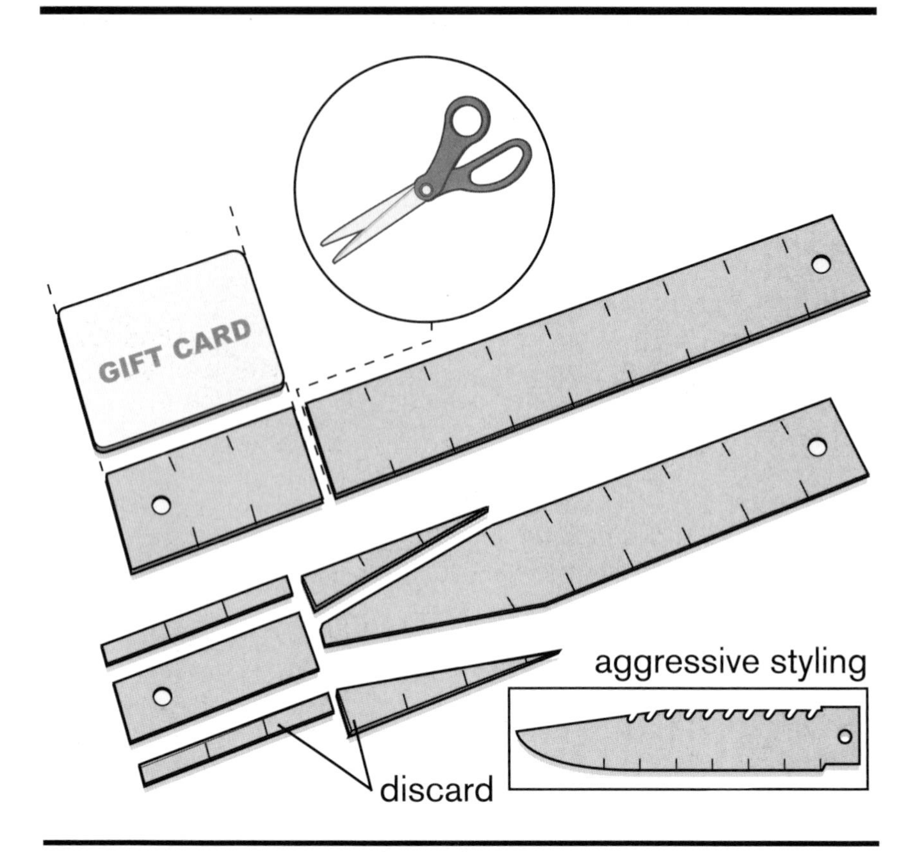

Place a gift card next to a flat plastic ruler, as shown. Use the gift card as a measuring template. Then with scissors, remove a gift card–long section from the end of the plastic ruler (top image).

Now, from the small section of ruler you just removed, cut off two 1/4-inch strips. Discard the center piece and save the two 1/4-inch strips for step 4.

At one end of the longer ruler section, use scissors to cut out a knife shape, as shown.

Step 2

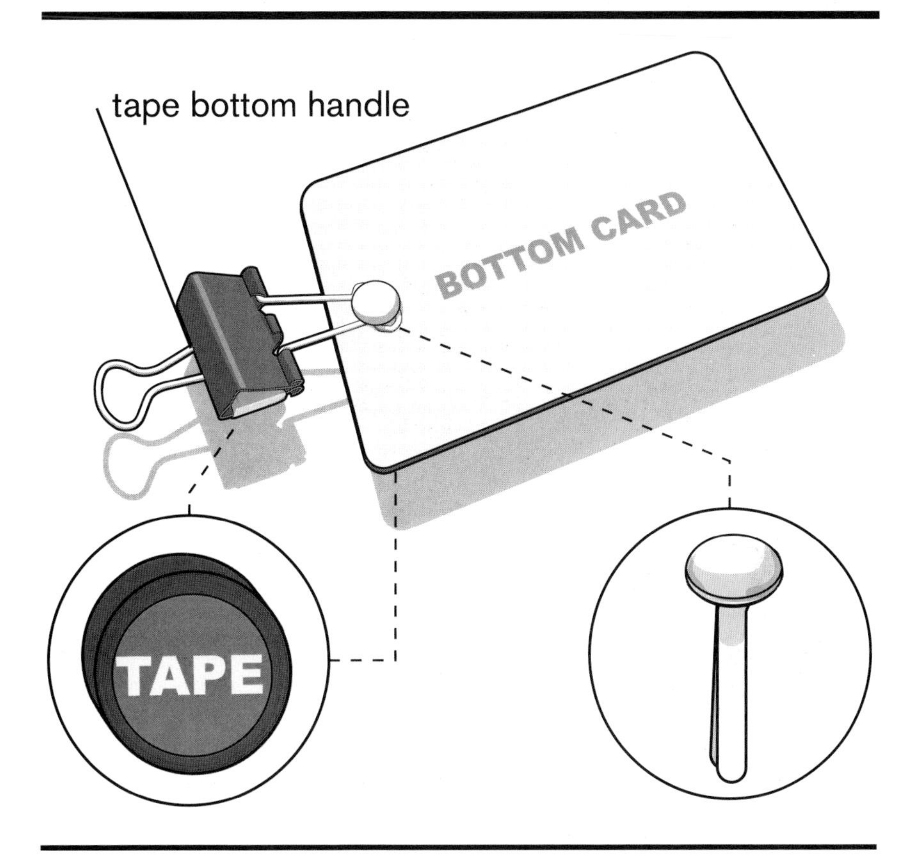

To control the spring-loaded ruler blade, you will need to lock the blade in place when not in action. To do this, use a brass fastener and a small binder clip.

First, duct tape or hot glue one of the binder clip's metal handles onto the binder clip housing. Then puncture or cut a hole about ½ inch from the edge of a gift card, centered as shown. (You may need scissors or a hobby knife to pierce the plastic card.)

Finally, slide a brass fastener through the free, unglued binder clip handle and into the gift card. Bend the fastener clips to lock the fastener into place. This completes the spring-loaded locking mechanism.

Step 3

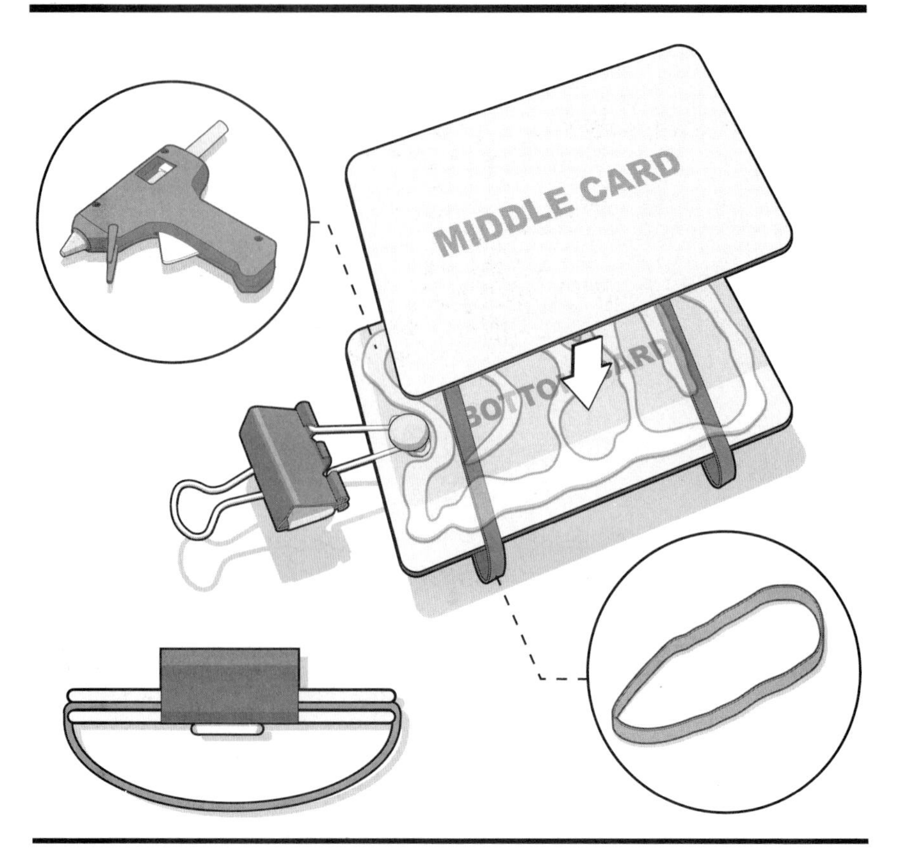

Test fit two of the rubber bands on your wrist, then remove them. If the bands fit comfortably, place both of them around the binder clip/gift card assembly as shown. Keep the bands far apart to maximize stability. Hot glue another gift card on top of this assembly, sandwiching both rubber bands and the brass fastener head between the two cards. Carefully hold everything in place as the glue cools.

This is the base of your Assassin Blade holder.

Step 4

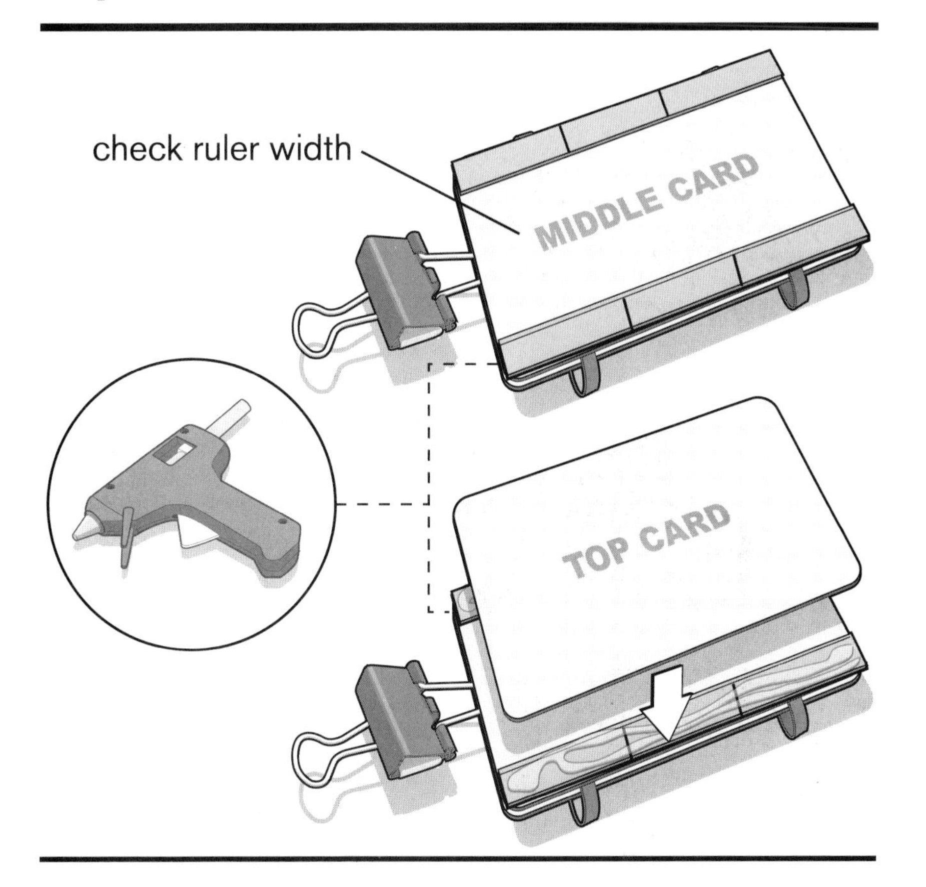

Check the ruler's width by "dry fitting" both ¼-inch plastic ruler slices on the face of the middle gift card (top image). The blade ruler should move easily down the middle channel. If not, reduce the side slices as needed.

If the ruler clearance is good, hot glue the ¼-inch slices onto the middle card as shown in the bottom image. Avoid glue overflow into the channel, as this will stop the blade from moving.

Once cooled, add hot glue on top of the two ¼-inch slices, then add the last card to the top. Again, be careful not to get glue in the channel.

Step 5

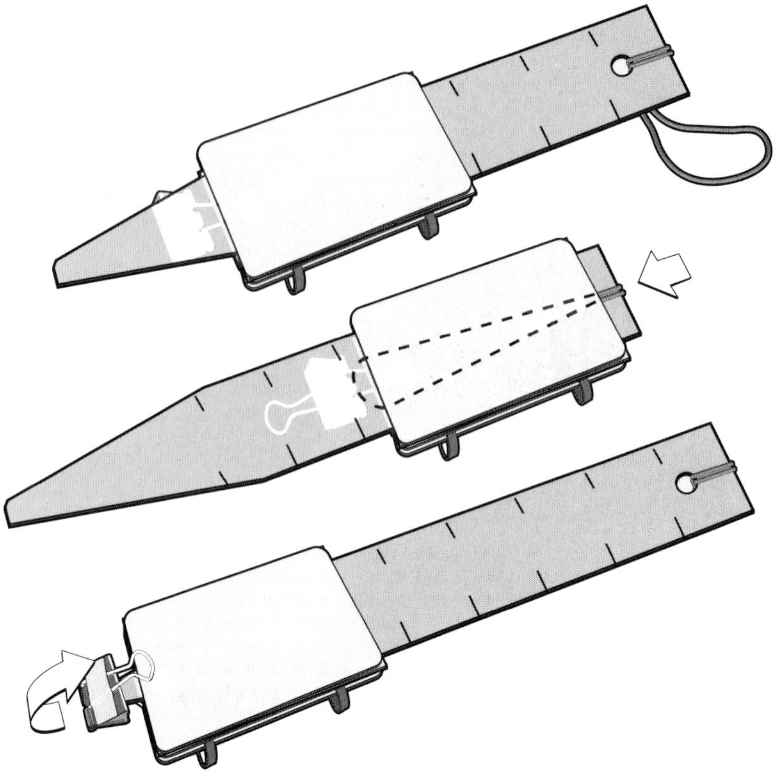

Now slide the plastic ruler blade into the channel, with the custom blade detail facing the binder clip. Tie a third rubber band to the back of the ruler, looping the band through the molded peg hook hole. Run the other end of the rubber band beneath the card and then wrap it around the attached binder clip (path shown with a dotted line). The ruler blade should now spring forward, tip out. If this is not the case, you will need a shorter rubber band.

Once adjusted, attach the whole assembly to your wrist. Slide the ruler blade back and flip up the binder clip to lock it in place, back up your sleeve, until the blade is needed. When you're ready to strike, flip down the clip and watch the blade spring out and stop, ready for action.

DETONATING PEN

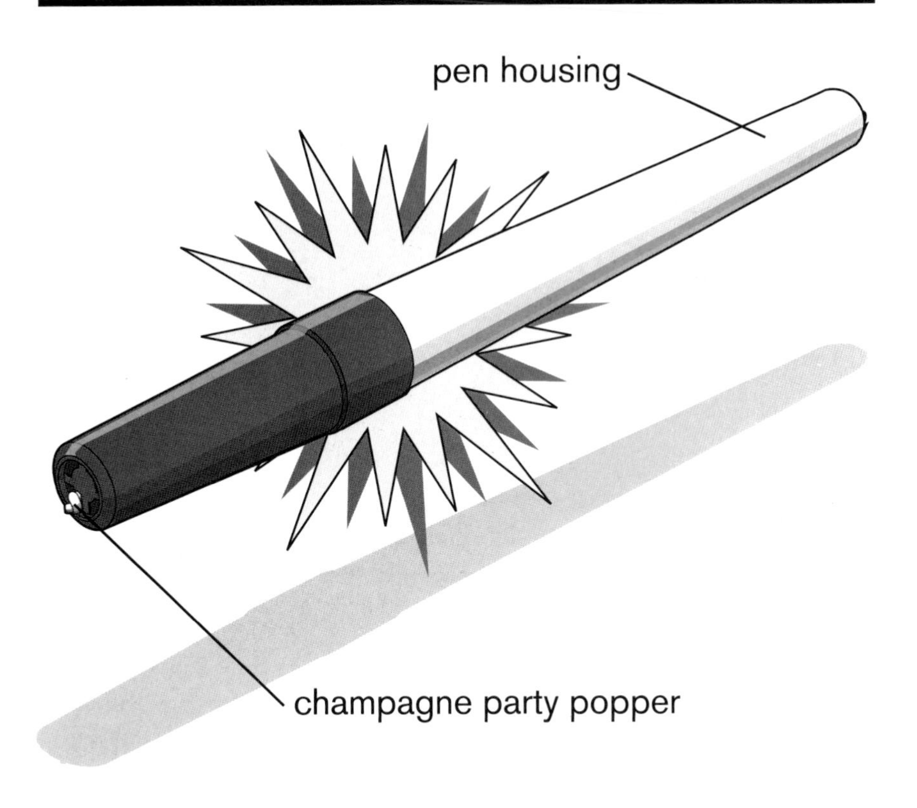

The Detonating Pen is designed to intentionally startle or surprise a clandestine target by emitting a loud popping noise when it's opened. The pen housing is packed with a nonlethal friction-actuated explosive charge (a party popper), which detonates when the cap is pulled off the pen. Though the gadget is nonflammable, it can cause ringing in the ears.

Supplies

1 champagne party popper
1 plastic ballpoint pen

Tools

Safety glasses
Glue
Scissors or hobby knife
Pushpin

Step 1

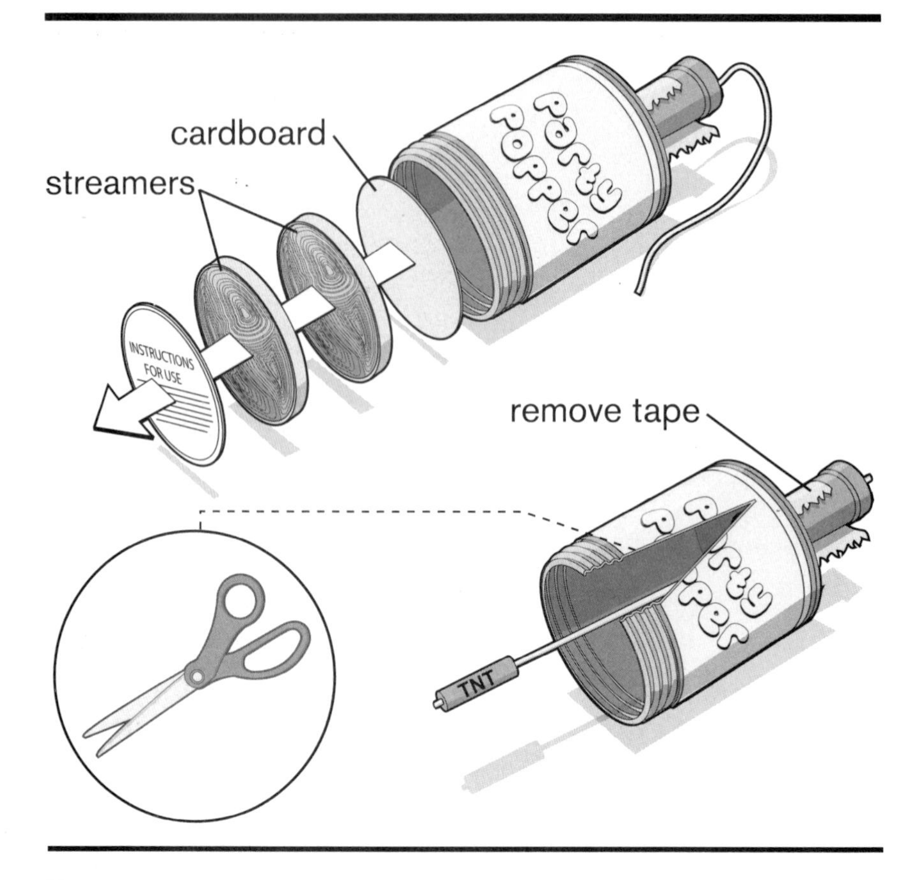

To start, you'll need a party popper that has a pull-string trigger. These party poppers are not considered fireworks and are available at most major retailers that carry party supplies. Use your finger or the tip of a pen to remove the streamers and cardboard. This will make it easier to locate the small explosive charge inside.

While wearing safety glasses, remove the tape or decorative foil around the back neck where the trigger string hangs out.

Using scissors, cut out a side section of the party popper from the front, as shown. This will provide room for your fingers to grab the explosive charge, which is housed in a cardboard tube. Once you have it, slowly slide it out with the string still attached. If the cardboard tube feels like it's not coming out, ***don't yank it***–this might detonate the explosive charge. Take your time, and when you're finished, discard the popper housing.

Step 2

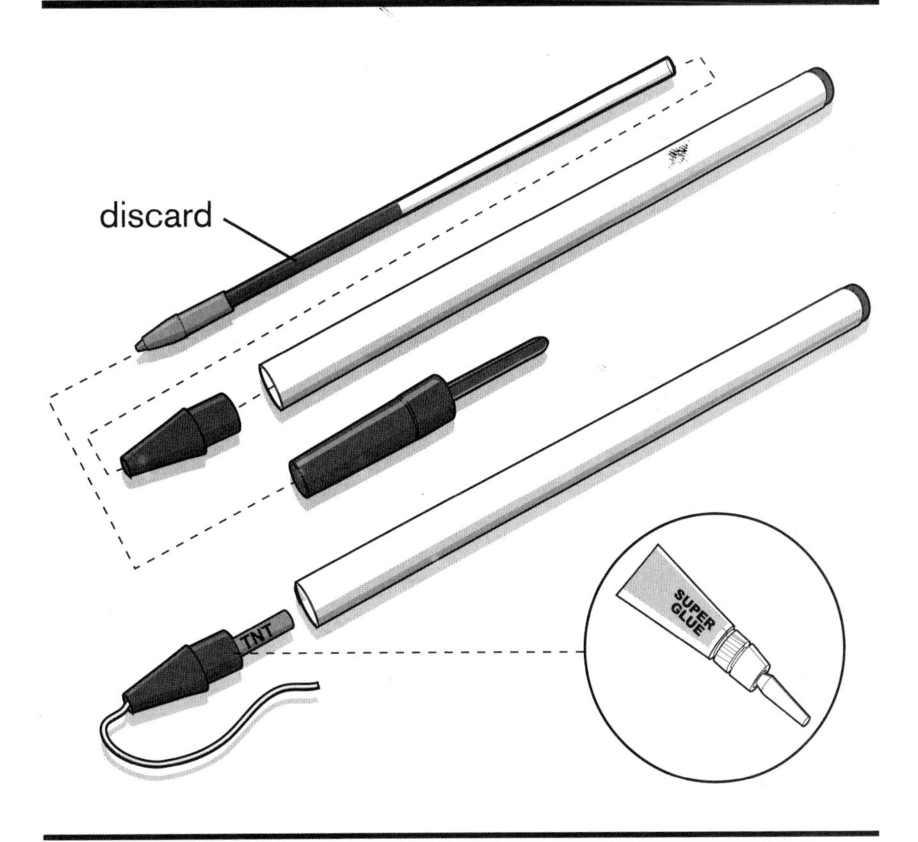

Disassemble a plastic ballpoint pen into its various parts, but do not remove the rear pen-housing cap.

Now slide the party popper trigger string through the plastic pen tip from the back, putting the small explosive charge in place of the ink cartridge.

If the pen tip does not screw onto the pen housing, dab a small amount of glue inside the pen tip to ensure that it will stay place. Adding glue will limit the Detonating Pen to one-time use.

Step 3

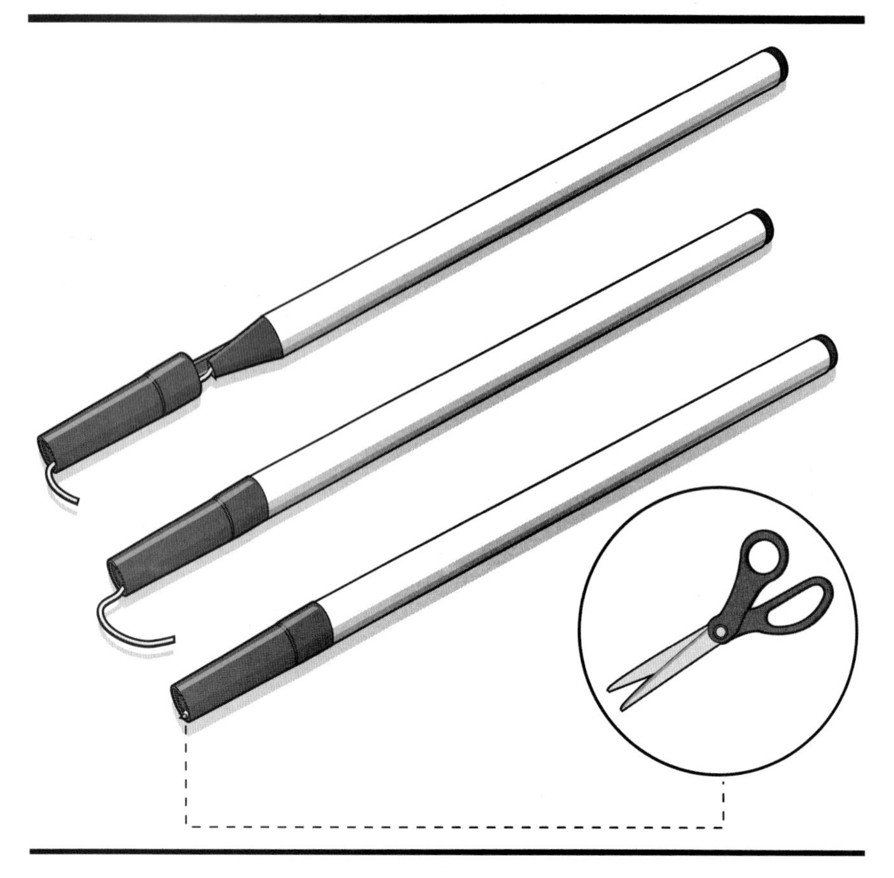

Snap the pen-tip cap back into the pen housing. Then run the party popper trigger string through the end of the original pen cap (top image). If the ballpoint pen cap does not have any air holes molded into its end, make one by carefully using a pushpin or hobby knife.

Once the string is through the cap, snap the cap onto the pen housing. Carefully tighten the string inside the pen cap and then tie a very small knot to keep it in place. (You may add a dab of glue onto the knot to make sure it holds.) Using scissors, cut off the excess string hanging out of the tip to make your booby trap less obvious.

Hold onto to this noise gadget until you need it. Does your intended target need something to write with? "Here, take my pen . . ."

SMOKING PEN

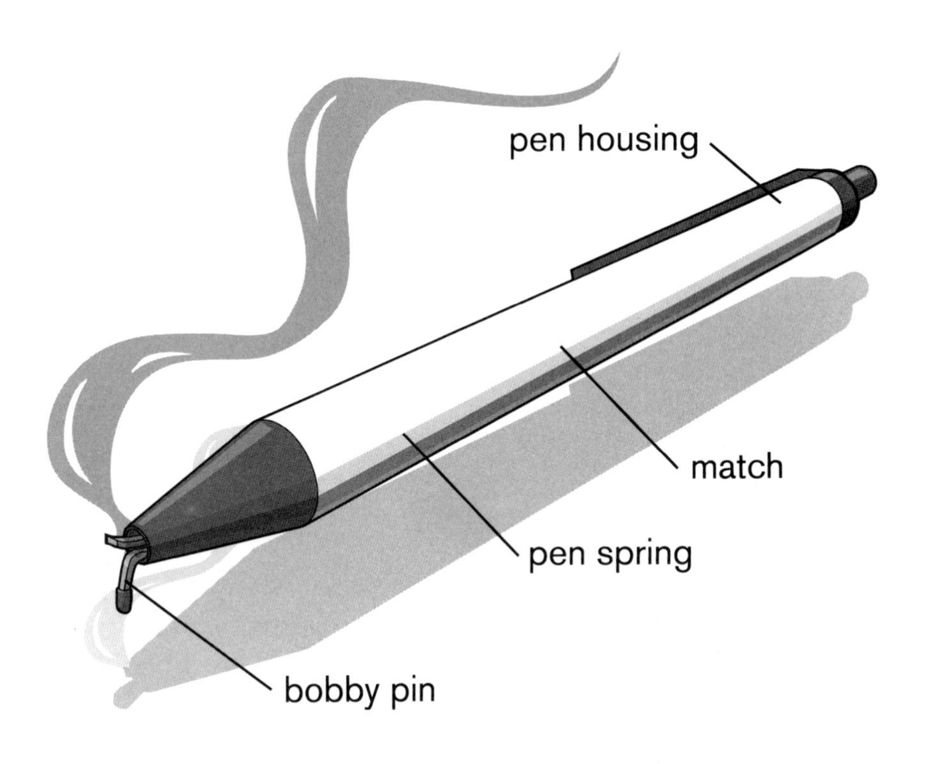

It might not be tear gas, but when activated this Smoking Pen will release a small cloud, a perfect distraction allowing you to slip out the back door. Its contents are simple: a strike-anywhere match and a modified bobby pin make the perfect getaway device.

Supplies

1 retractable pen
1 bobby pin
1 strike-anywhere match

Tools

Safety glasses
Pliers

Step 1

Disassemble a retractable pen by removing the ink cartridge, plastic tip, and small spring. Save the spring and tip, but discard the ink cartridge.

Slide a single strike-anywhere match into the pen housing, with the match head facing out. Shake the pen housing until the match falls to the bottom of the housing, near the clip.

Step 2

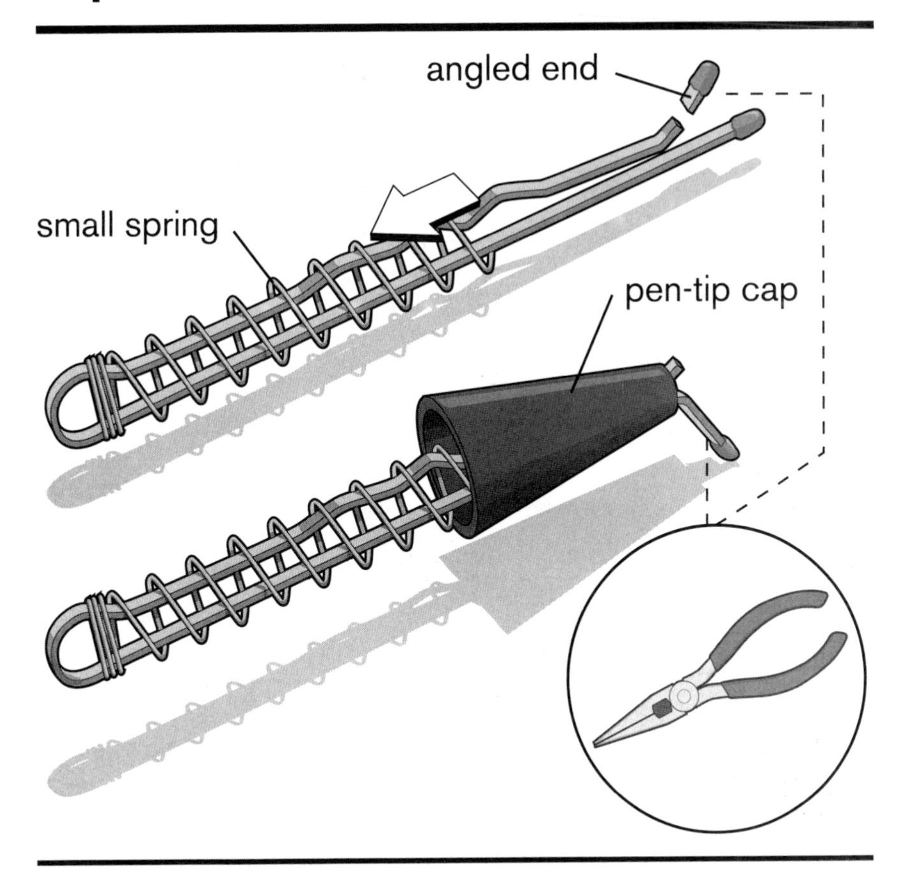

If necessary, use pliers to remove the angled end of the bobby pin, then slide the spring over both bobby pin bars until it reaches the circular end, as shown in the top image. Once in place, the spring should not be able to slide off the back; adjust the bobby pin or spring with pliers, if needed.

Now slide the pen-tip cap onto the bobby pin. Bend the uncut end of the bobby pin with pliers to lock the pen tip in place.

Step 3

Reassemble the pen tip assembly onto the pen housing as shown above. When you need to activate the smoke screen, pull the bobby pin forward and release it. The bobby pin will spring back, smashing into the strike-anywhere match head, creating heat from the friction. This heat will ignite the match and release a stream of smoke from the tip of the pen.

Before you discard or store this MiniWeapon, make sure the match inside has stopped burning.

5
SURVEILLANCE AND INTEL

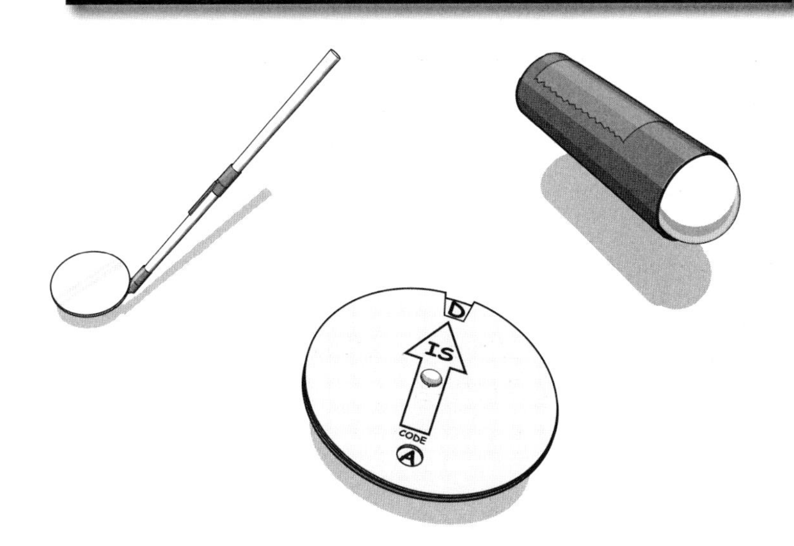

BIONIC EAR

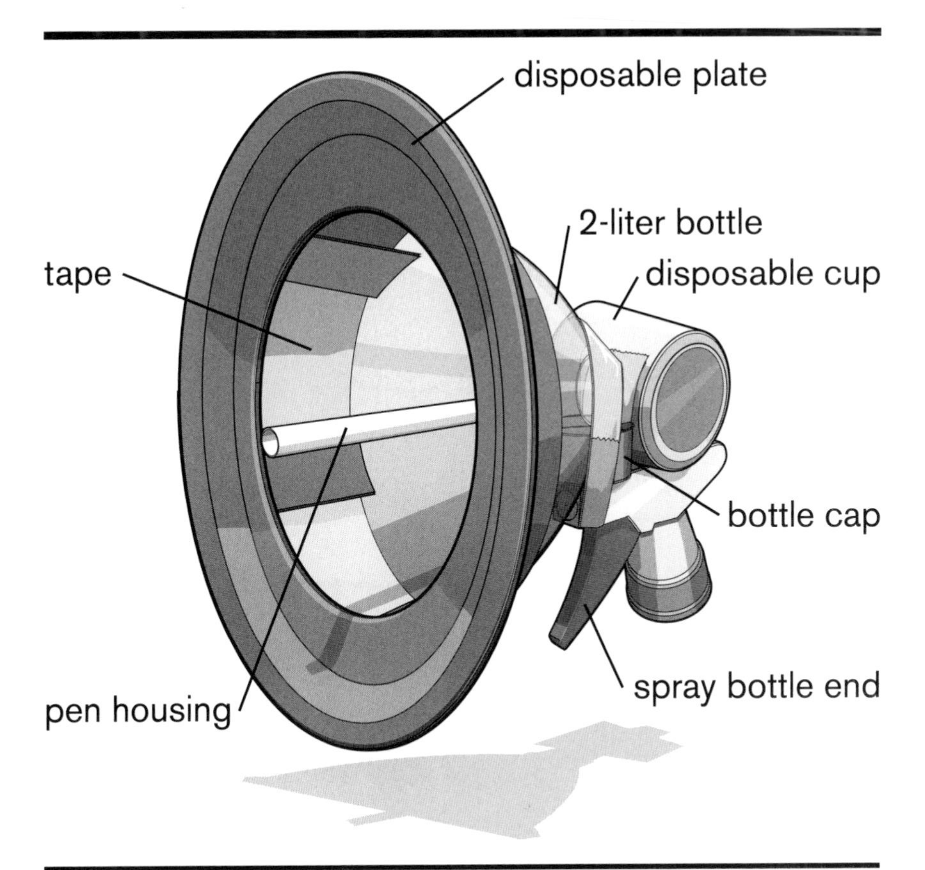

Simply point this Bionic Ear at your subject to capture secret discussions. Its curved dish will focus and funnel whatever information you are trying to steal into an earpiece that fits perfectly around your ear. This is a must for any stakeout.

Supplies

1 2-liter plastic soft drink bottle
1 disposable plastic or paper plate (greater than 7 inches in diameter)
Duct tape
1 plastic or paper disposable cup
1 spray bottle nozzle

Tools

Hobby knife
Marker
Scissors
Hot glue gun

Step 1

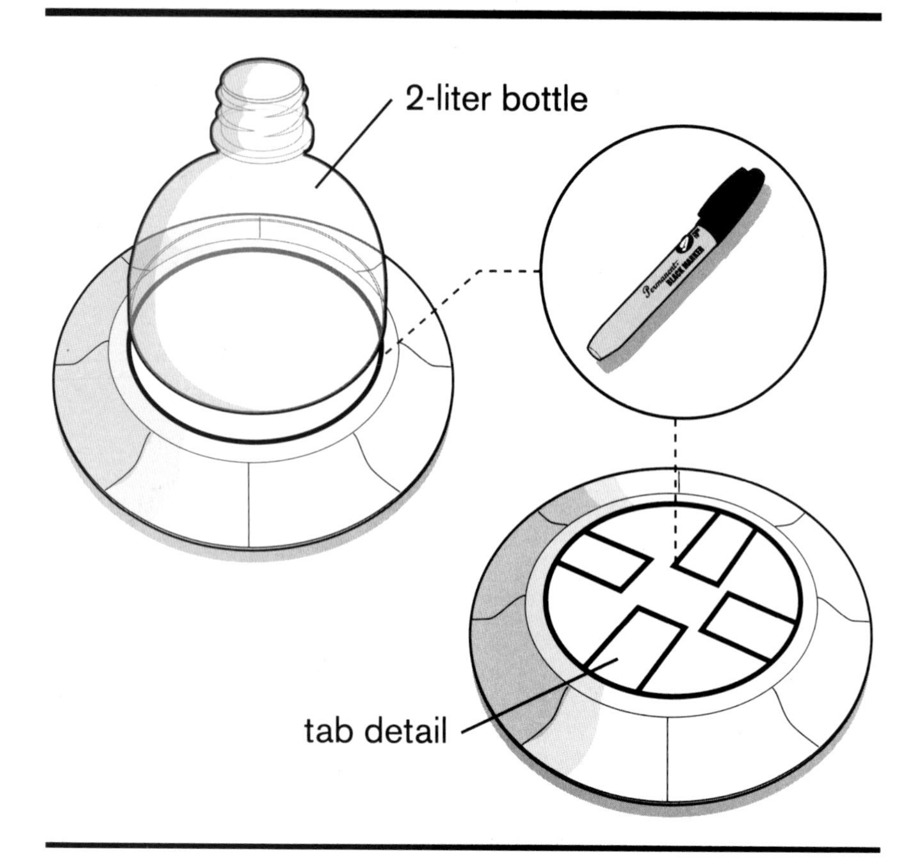

Using a hobby knife, remove the top 4 inches of a 2-liter plastic soft drink bottle. Save the plastic bottle cap for step 4 but discard the base of the bottle.

Place the top of the bottle onto the disposable plate and trace the bottle's diameter onto the plate as shown. Remove the bottle and draw four tab details onto the plate within the marked circle. These tabs will be used later to secure the disposable plate to the bottle.

Step 2

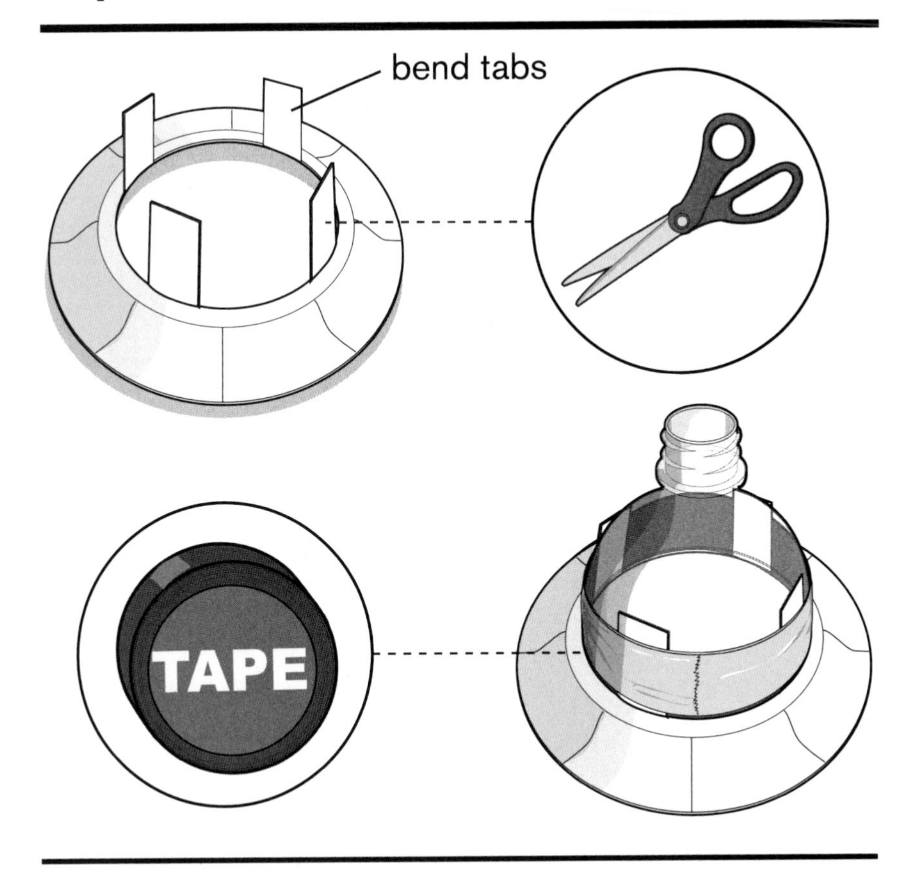

Keeping the drawn tabs attached to the outer rim, use scissors to remove the surrounding area *within* the drawn circle on the plate surface. Once you have removed the material, bend the tabs toward the back of the plate as shown.

Next, slide the 2-liter bottle top in between the tabs and use duct tape to secure it in place. You have just completed the dish that catches the audio.

Step 3

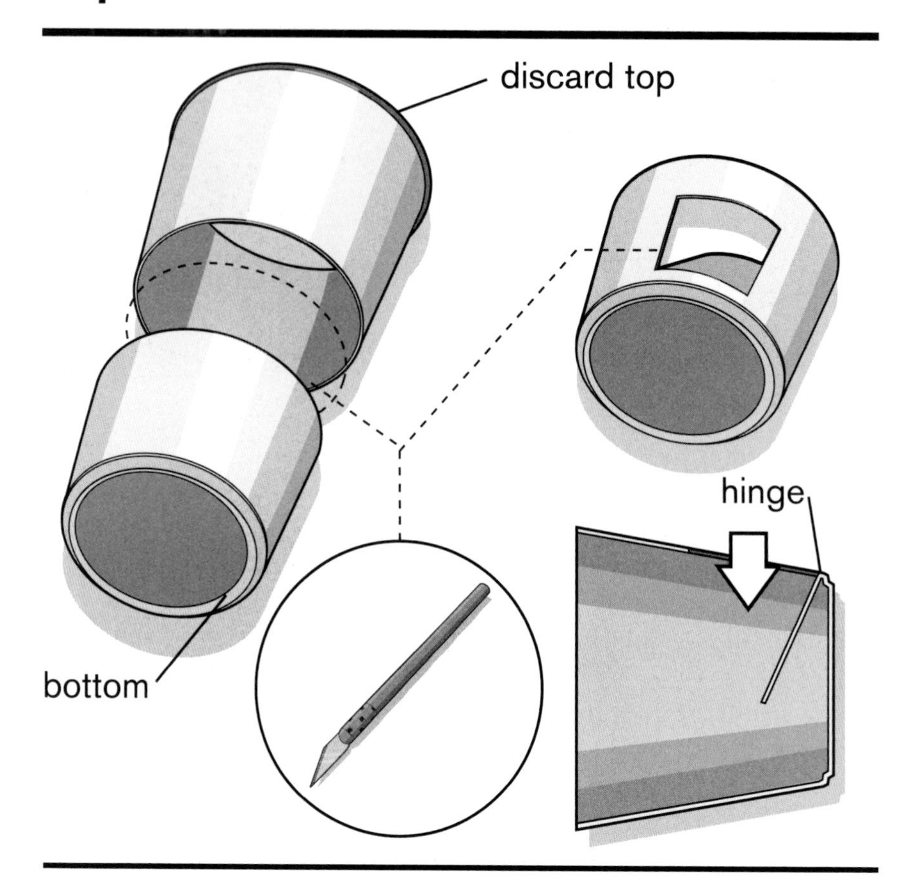

Use a hobby knife or scissors to remove the bottom 2 inches of the plastic or paper disposable cup. Discard the top of the cup and keep the base.

Now, on the saved base section, slice out a 1-inch-by-1-inch trap door as shown. The end of the 2-liter bottle will eventually fit into this hole. Cut only three of the four sides, then hinge the door at a 45-degree angle. When the Bionic Ear is completed, sound will bounce off this angled flap and into your ear.

Step 4

A nozzle from a spray bottle will make a suitable frame and handle for this homemade Bionic Ear. Twist off the nozzle from an old spray bottle and clean out any leftover liquid still trapped inside the nozzle.

Hot glue the 2-liter bottle cap upside down on the top of the spray nozzle, near the dispenser end. Let it cool. Then hot glue the modified cup bottom to the back end of the spray nozzle. The 1-inch-by-1-inch hole you cut out should be in line with the bottle cap. See the illustration.

Step 5

Slide the custom listening dish (2-liter bottle) on top of the glued bottle cap. The bottle neck should insert into the hole in the modified cup. Once in place, fasten it using duct tape. Next, use tape to seal all the holes so that sound cannot escape from the sides but is instead funneled into the cup. Aim the Bionic Ear toward sounds and then slide the cup over your ear. The device should amplify faint sounds up to 20 feet away, though results will vary.

Professional Bionic Ears feature a built-in antenna protruding from the dish. You can add one (for visual effect) by hot gluing an empty pen housing to the inside of the 2-liter bottle end. It will look something like the finished illustration on page 189.

TOOTHPASTE PERISCOPE

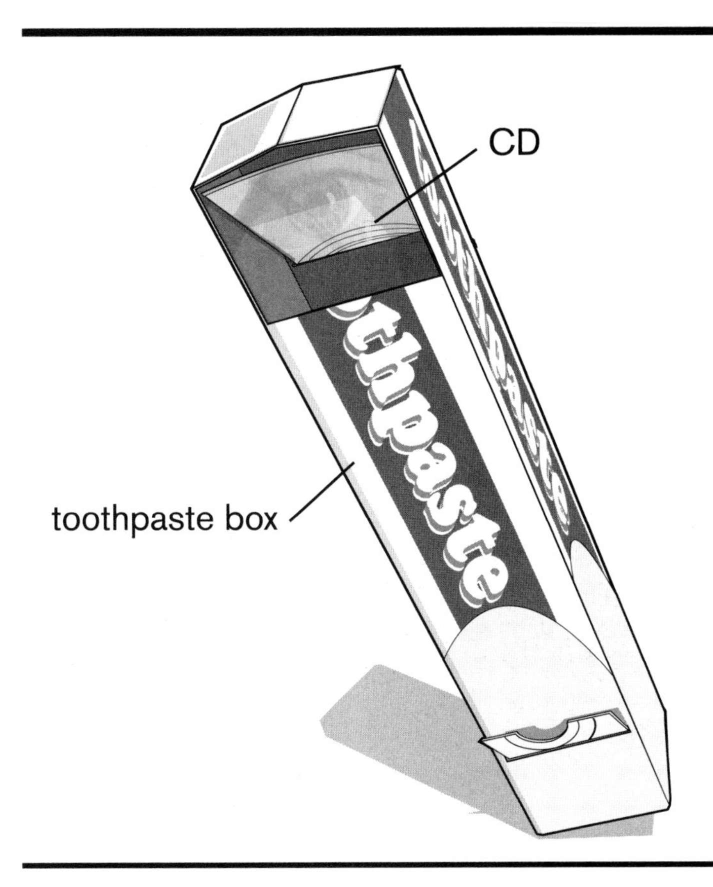

As a spy, you must keep your true identity a mystery. Observing and obtaining information from a concealed location will accomplish this. To avoid being detected, you need the ingenuity of a good old-fashioned periscope. The Toothpaste Periscope is constructed with two simple CD "mirrors" and a large toothpaste box. It will allow you to see around corners or over walls, yet keep your head out of sight!

Supplies

1 toothpaste box
1 unwanted CD or DVD
Clear tape

Tools

Marker
Scissors
Hobby knife

Step 1

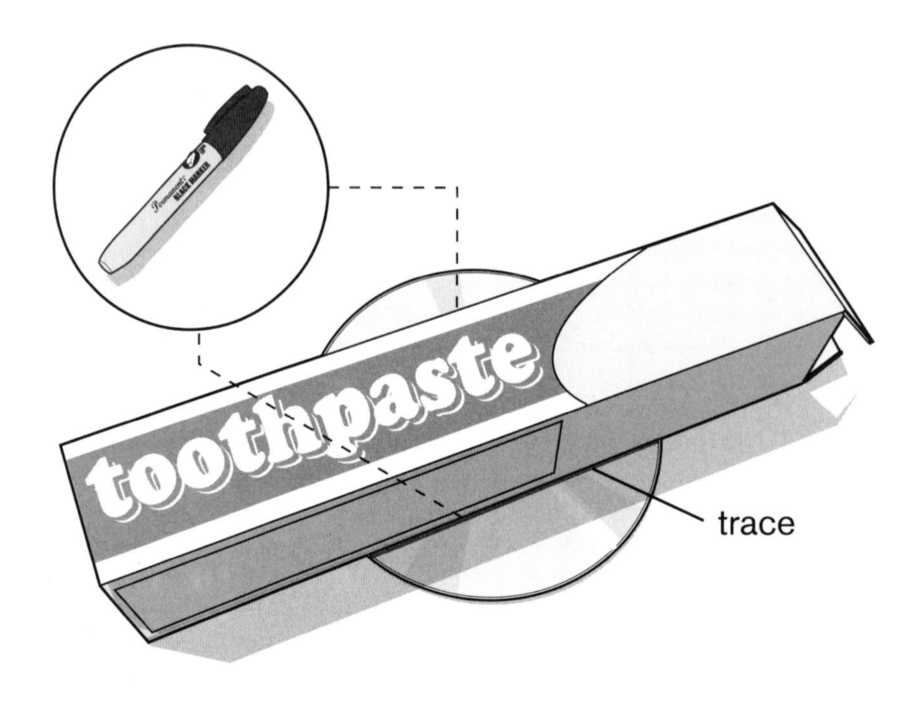

Center an empty toothpaste box on a highly reflective unwanted CD or DVD as shown. Then, use a marker to trace the box's outline (width) onto the disc surface.

Step 2

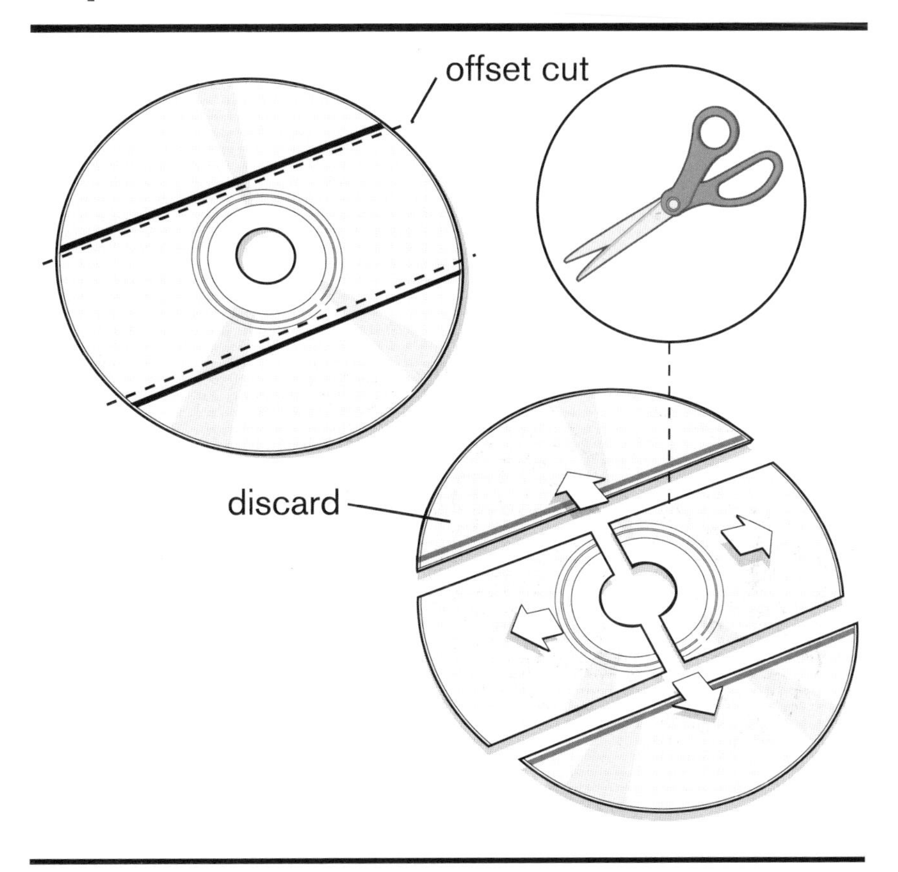

Now cut two small mirrors from the disc. Using scissors, cut the disc slightly narrower than the drawn lines so that it will fit inside the toothpaste box.

When the disc is in three individual sections, use the scissors to cut across the center section, creating two halves, as shown in the bottom image.

Discard the two long sides; you will only be using the two center segments.

Step 3

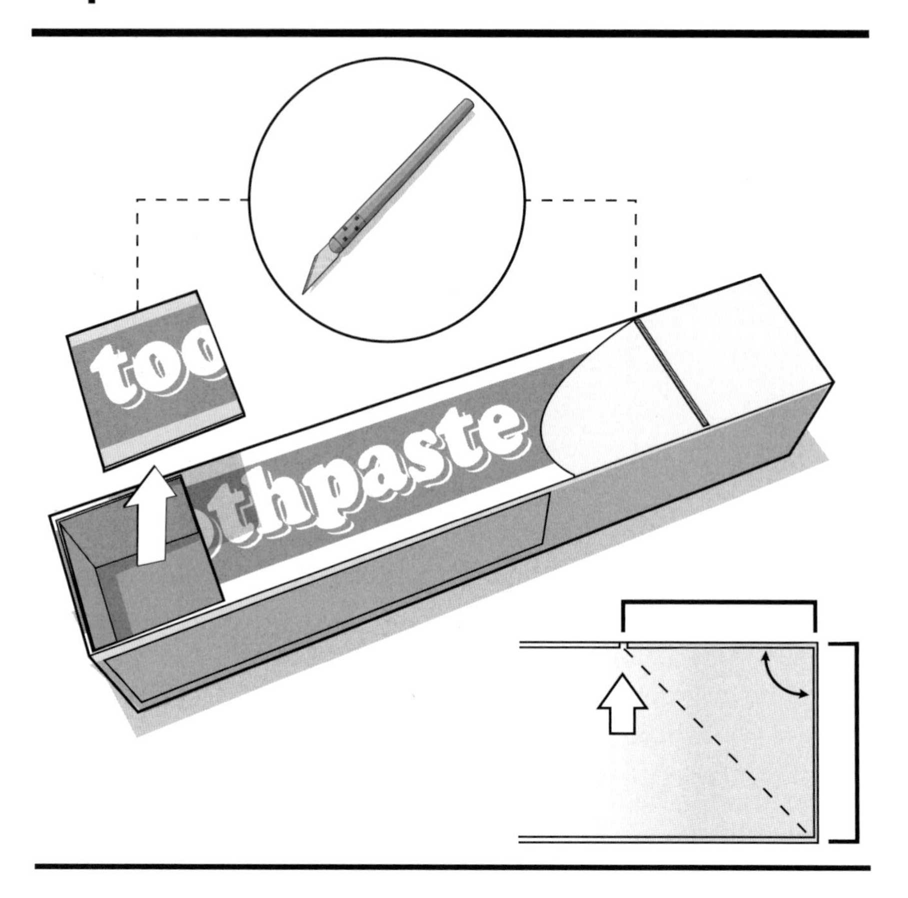

Modify the empty toothpaste box to create eye holes and mirror slits. Carefully cut out and remove a square opening approximately 1 ½ inches from one end of the toothpaste box. On the same side, cut a small slit approximately 1 ½ inches from the opposite end.

Step 4

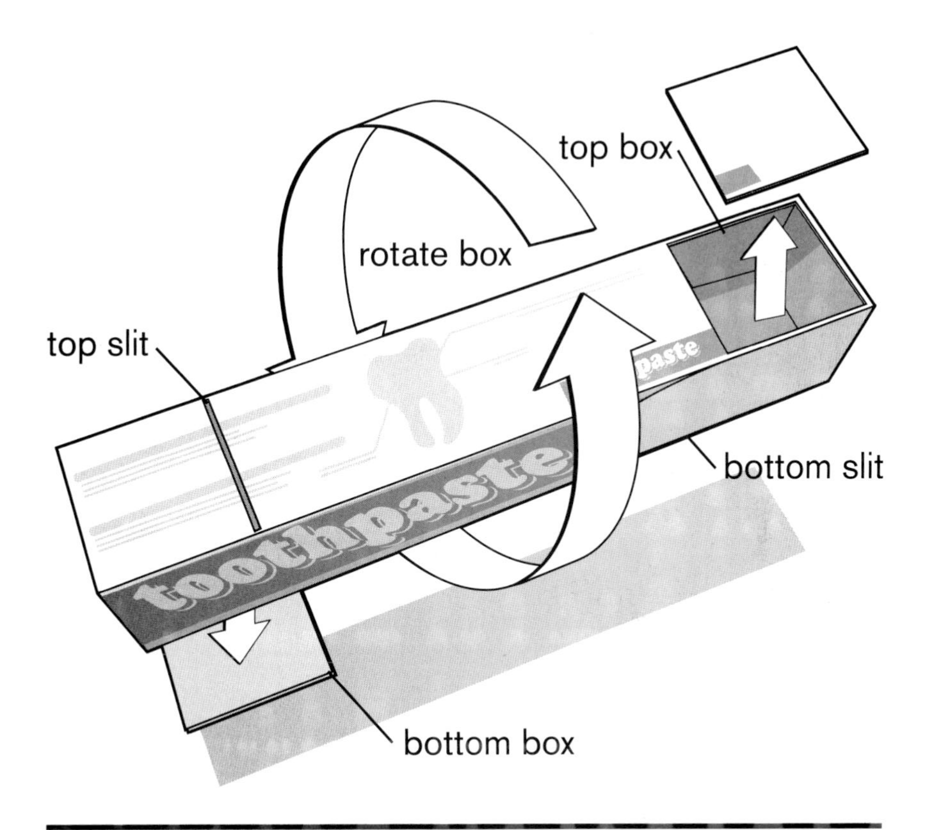

Flip over the box and repeat the previous step, but mirrored. On the reverse side of the end with the original small slit, cut out and remove a square opening approximately 1 1/2 inches from the end. On the same side, but opposite end (with the first eye hole beneath it), cut a small slit approximately 1 1/2 inches from that end, as shown.

Step 5

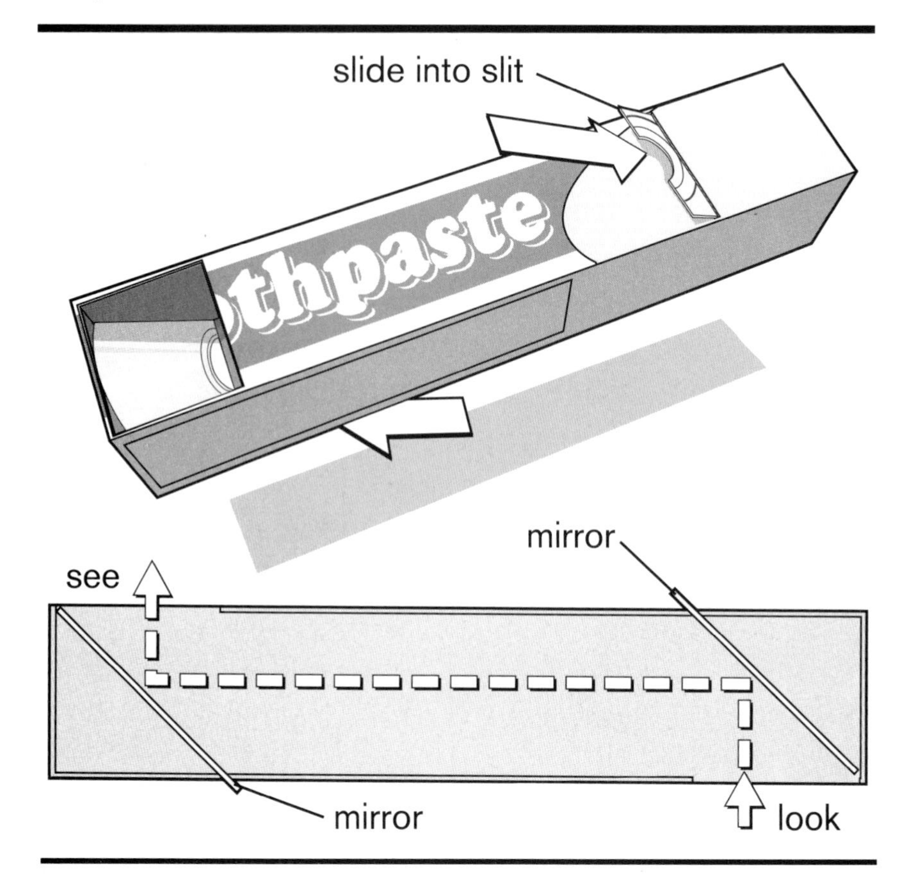

Slide a single disc mirror through each slit at a 45-degree angle. Both mirrors should be angled parallel to each other with their reflective surfaces facing into the box. Hold your eye up to one eye hole and test out the arrangement. Fasten the mirrors with clear tape if needed. (For better results, use two small health and beauty mirrors.)

MINI INSPECTION MIRROR

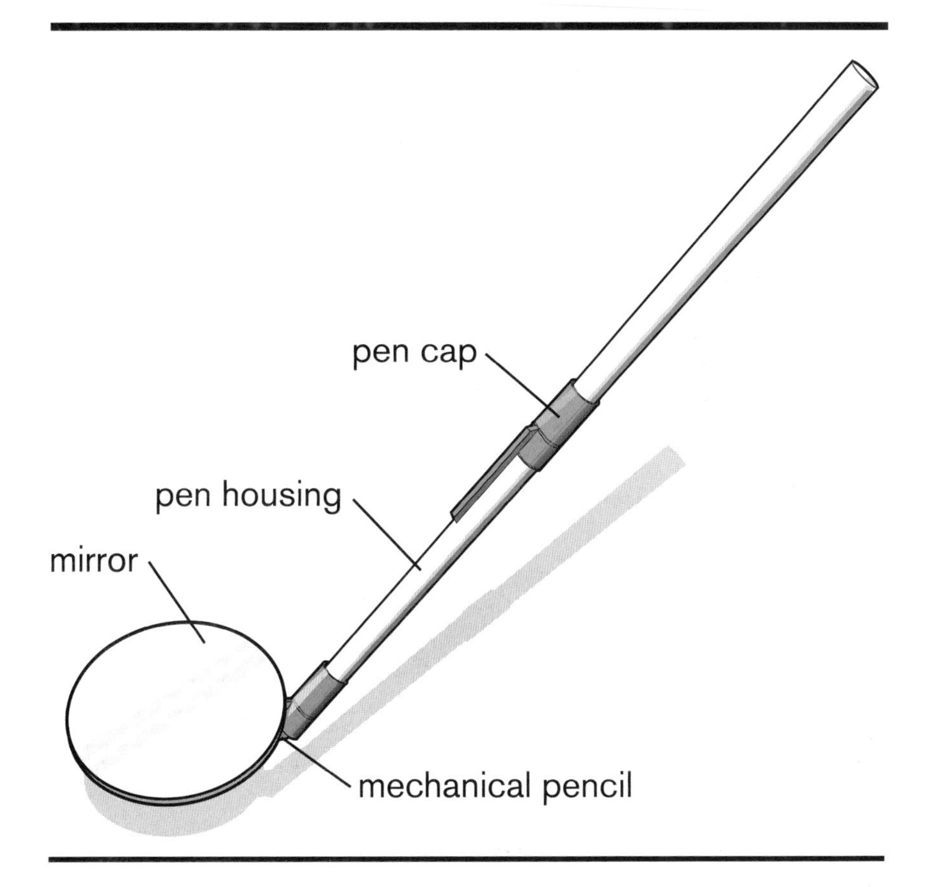

Whether you're sweeping a room for electronic bugs, looking for lost cash, or double-checking for hidden assailants, the Mini Inspection Mirror is perfect for viewing any area that is obstructed from your direct line of sight. It's a great tool, because a good spy should always be aware of his or her surroundings . . . including under the couch.

Supplies

2 plastic ballpoint pens
1 inexpensive mechanical pencil
1 makeup compact with mirror

Tools

Hobby knife
Pliers
Hot glue gun

Step 1

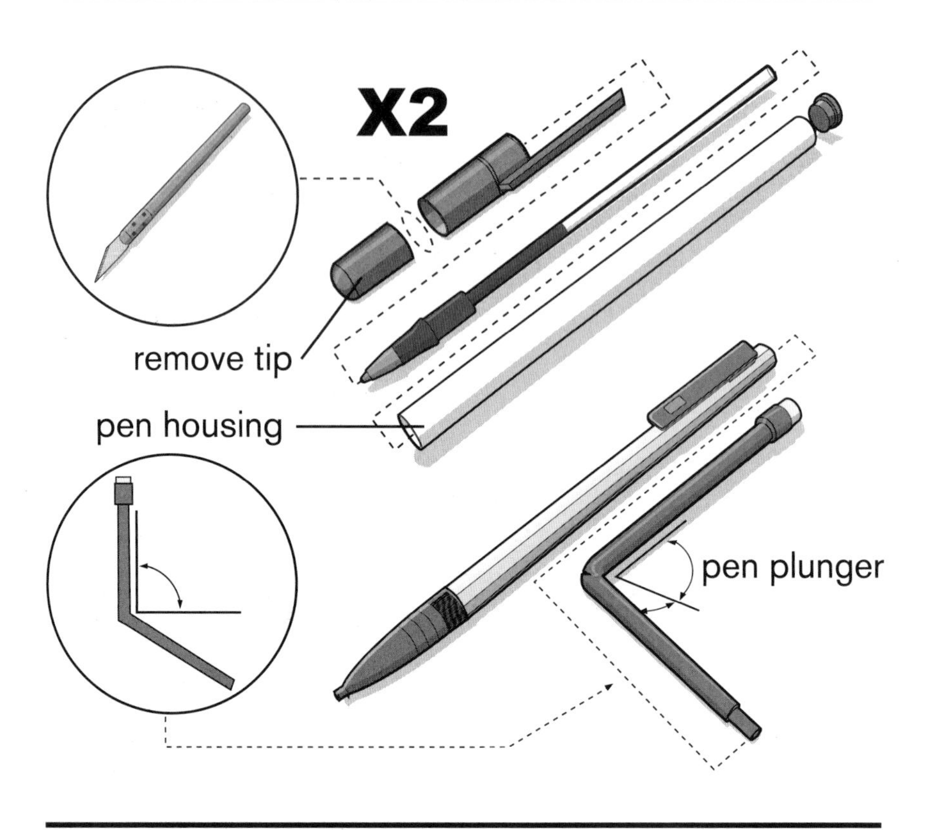

Completely disassemble two plastic ballpoint pens. You may need a hobby knife or pliers to remove the rear pen caps. Save these rear pen caps for making the Round-Nosed "Bullet" (page 73).

Using a hobby knife, cut both plastic pen caps in half. These will be used later to connect the various parts of the Mini Inspection Mirror. Next, remove the plunger from a mechanical pencil. Once removed, carefully bend the plunger at a 110-degree angle. Use the illustration to approximate the bend.

Step 2

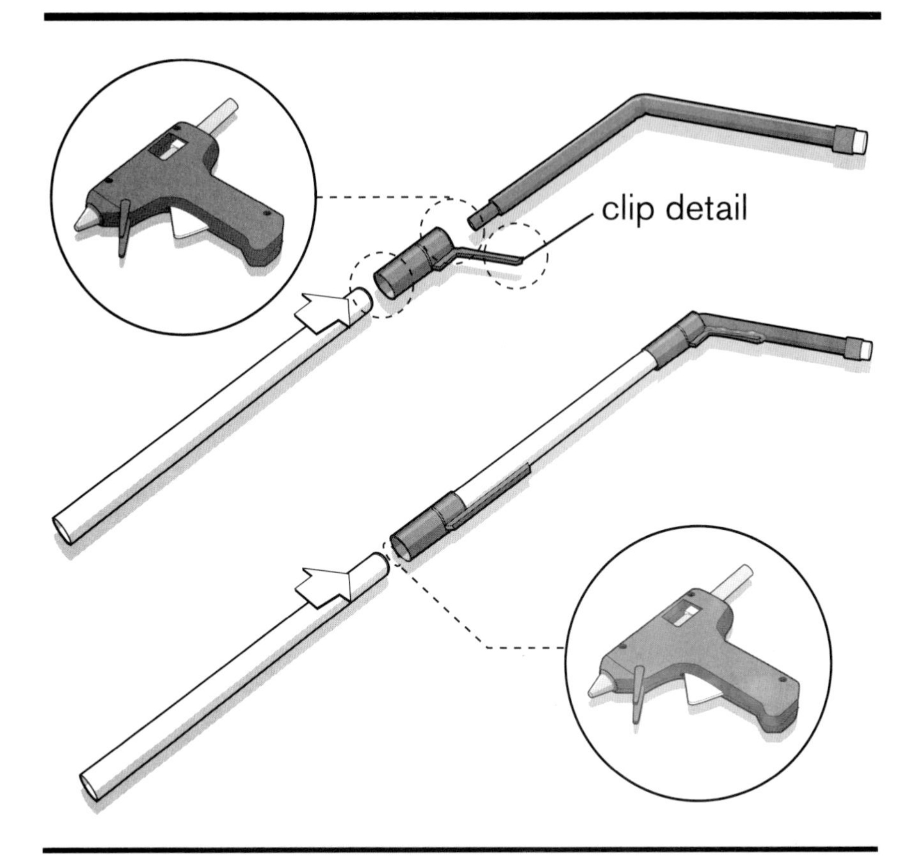

Hot glue one of the modified pen caps to the end of one pen housing, with the clip detail extending from the end, not overlapping the housing. Then slide the bent pencil plunger into the pen housing. Bend the pen clip to align it with the pencil plunger and push them together. Hot glue the clip to the plunger.

Finally, slide the second modified cap onto the end of the first pen, then connect it to the second pen housing. Hot glue these pieces together.

Step 3

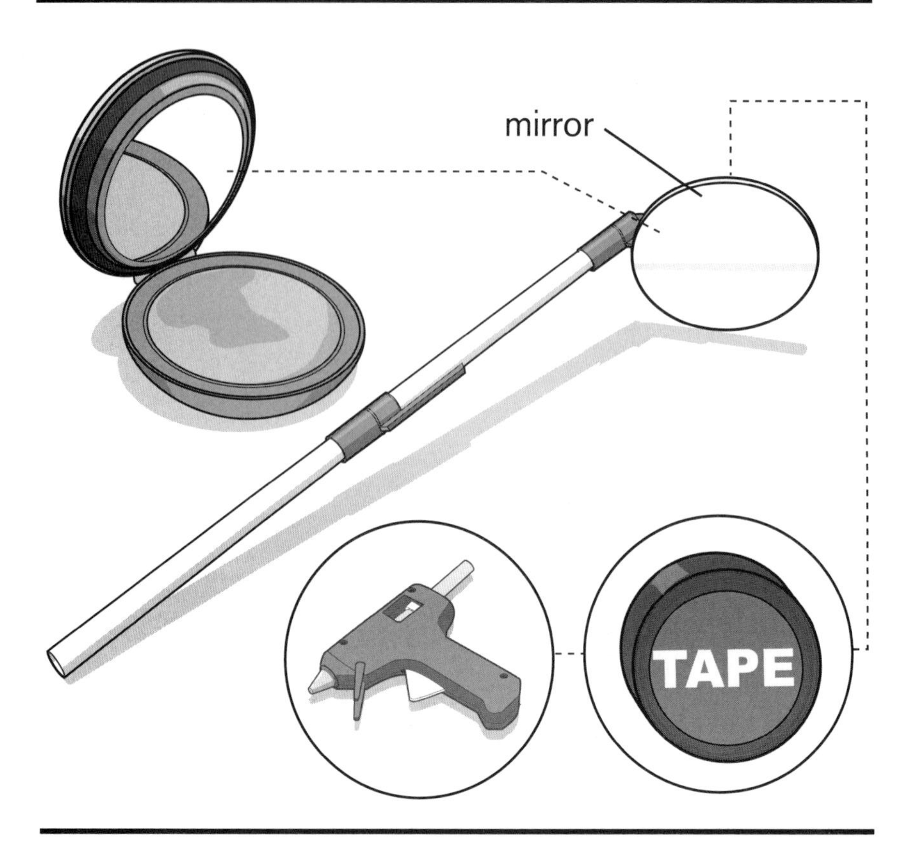

Remove the circular or rectangular mirror from the makeup compact using your own strength or pliers; this might be difficult depending on how well it's put together. Once the mirror is removed, use hot glue to attach it to the top of the pencil plunger. Holding the mirror in place, continue to secure it with duct tape.

If you do not have a makeup compact mirror, a highly reflective DVD or CD cut to size will also work. (See the Toothpaste Periscope, page 195.)

CODE WHEEL

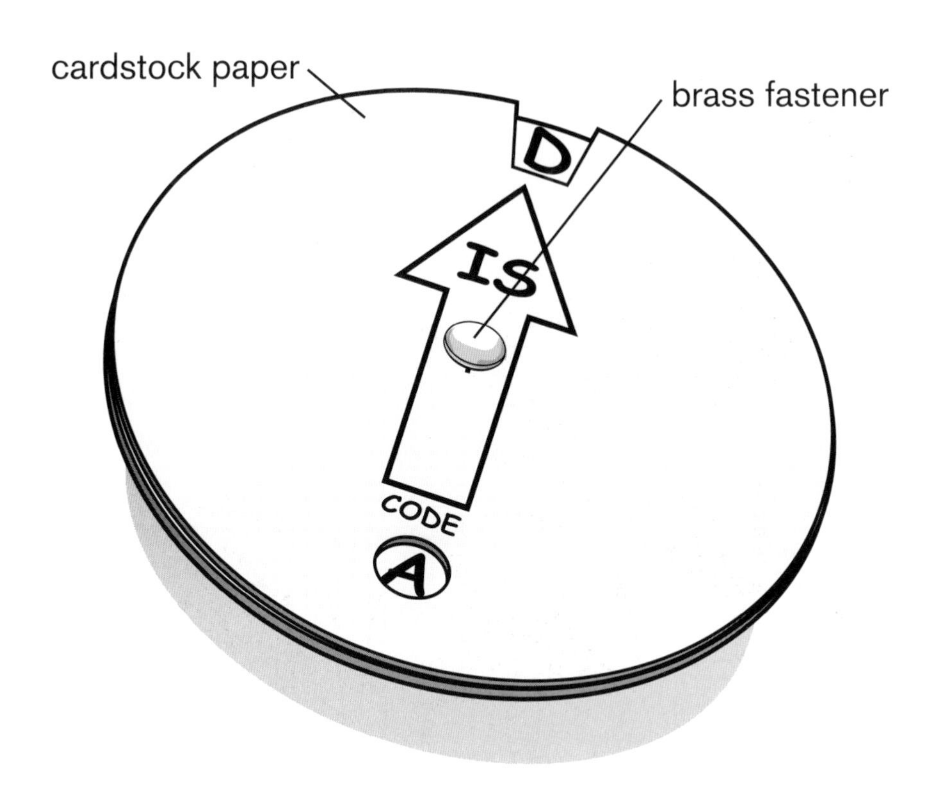

Need to convert a secret message into code? One of a spy's goals is to send and receive information, keeping it safe from anyone who tries to intercept it. This two-piece Code Wheel can be used to convert information or add observations to your secret mission journal or diary. This cipher wheel is constructed from two paper circles; when you twist them, different letters line up, slowly revealing an important message. We suggest you make two Code Wheels, one for the sender and the other for the recipient . . . just keep them out of enemy hands!

Supplies

1 8½-inch-by-11-inch piece of cardstock or cereal box
1 brass fastener
Duct tape

Tools

Marker
Unwanted CD or DVD
Scissors
Hobby knife or pushpin

Step 1

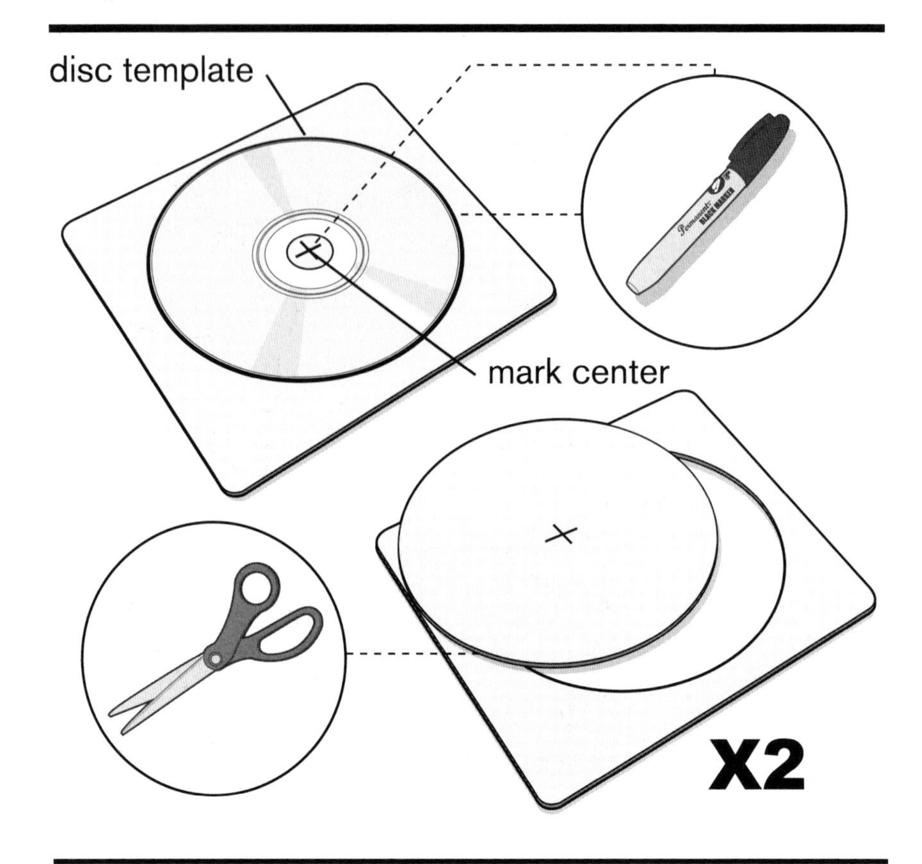

The individual wheels are constructed from one 8½-inch-by-11-inch sheet of cardstock. If this material is not available, use a cereal box or similar thin cardboard.

With a marker, trace two DVD or CD disc circles onto the cardstock. Before removing the disc, mark the center points of both circles with an "X."

Cut out both circles with scissors and discard the scraps.

Step 2

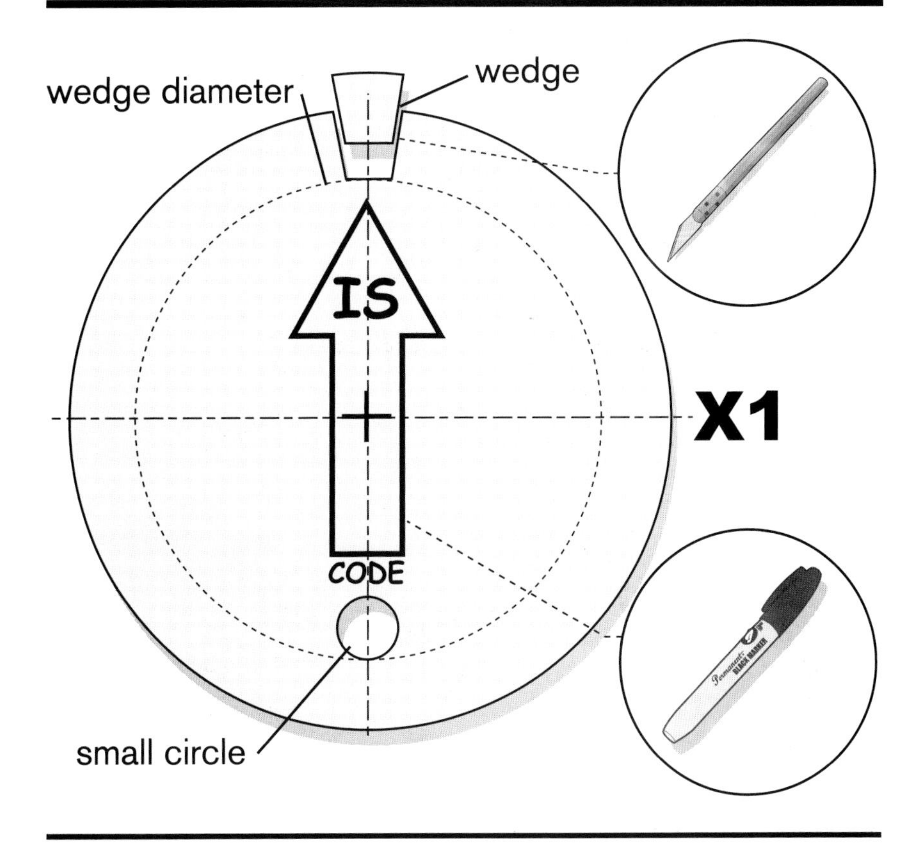

Now modify one of the paper discs. Use scissors to cut out a small wedge, ¾ inch deep and less than ½ inch wide.

Then, below the wedge diameter (illustrated with dotted lines), on the opposite end of the circle, cut out a small circle, no bigger than ½ inch in diameter.

Optional: draw a large arrow on top of the wheel and add the word "IS" at the point end and "CODE" below it. See the illustration.

Step 3

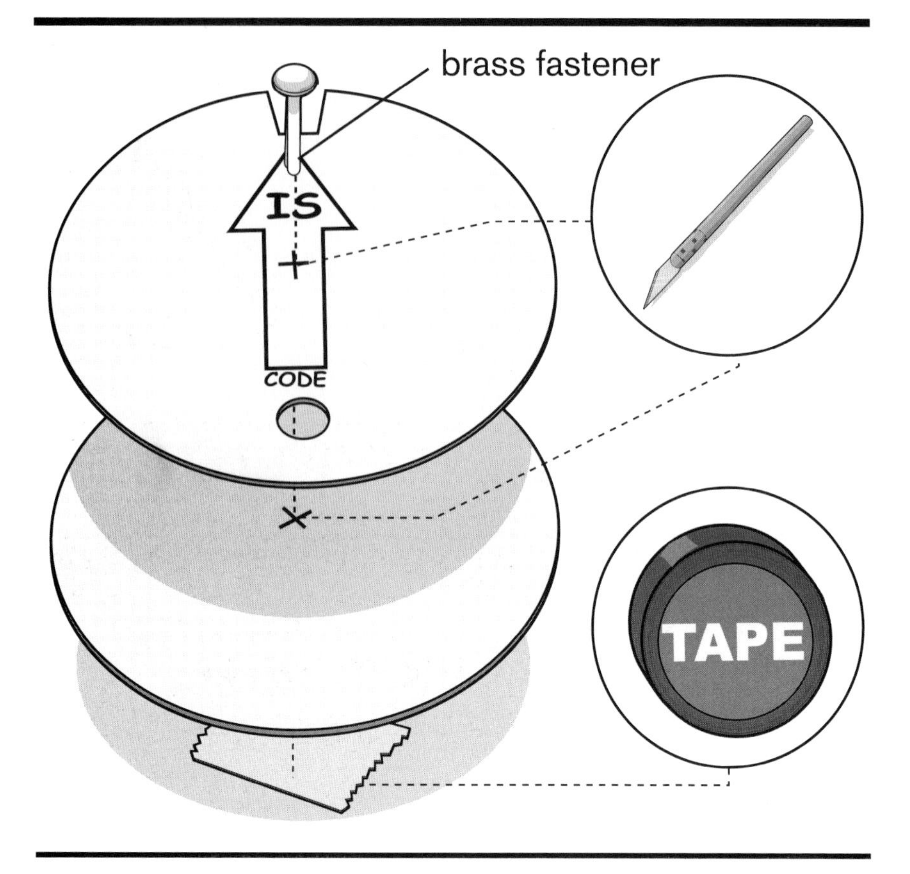

Poke two small holes in the center of both paper wheels using a hobby knife or pushpin. Line up the wheels and fasten them with a brass fastener. Add some duct tape over the fastener arms on the bottom side of the lower wheel to hold the entire assembly in place.

Step 4

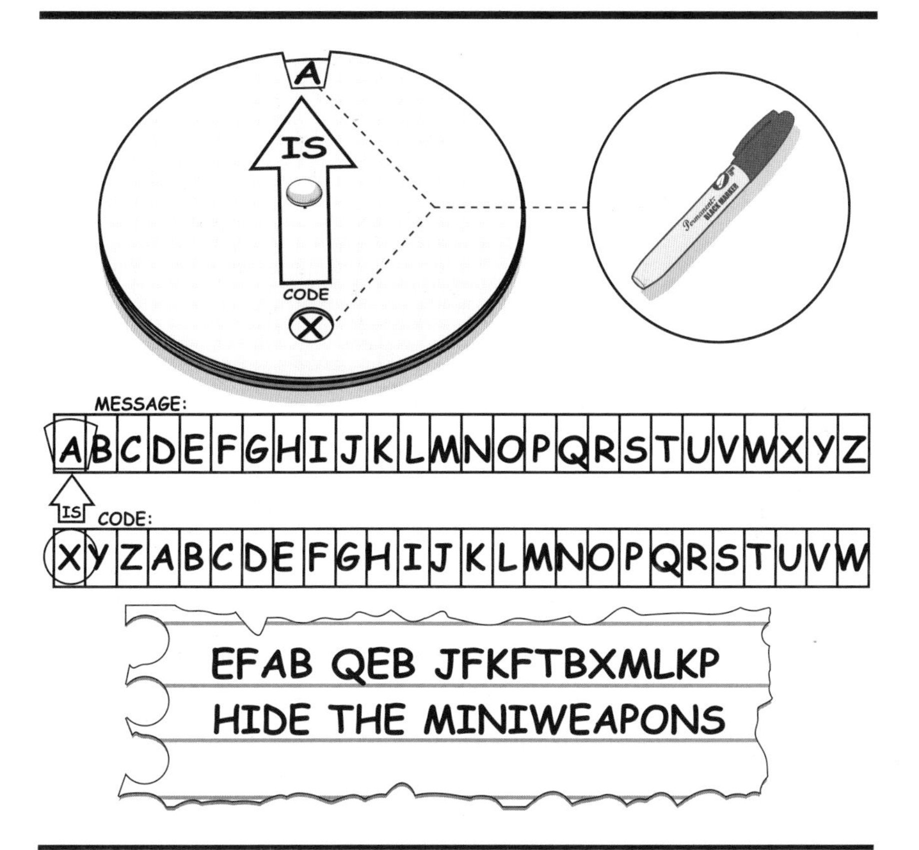

Between the cut wedge write the letter "A," then rotate the top wheel until that letter is out of sight. Now write "B." Continue turning and adding letters until you've completed the alphabet and numbers 0 to 9.

Next, turn the top dial back to "A" and hold the wheel in place. In the small circle opening, write the code letter you wish "A" to stand for. In the case above, we've written "X." Continue adding unused letters and numbers (keep track!) until you've completed the alphabet and numbers 0 to 9.

To send or receive a message, both parties involved must have identical Code Wheels, so build another using the same letter-to-letter code as the first.

Alternate Construction

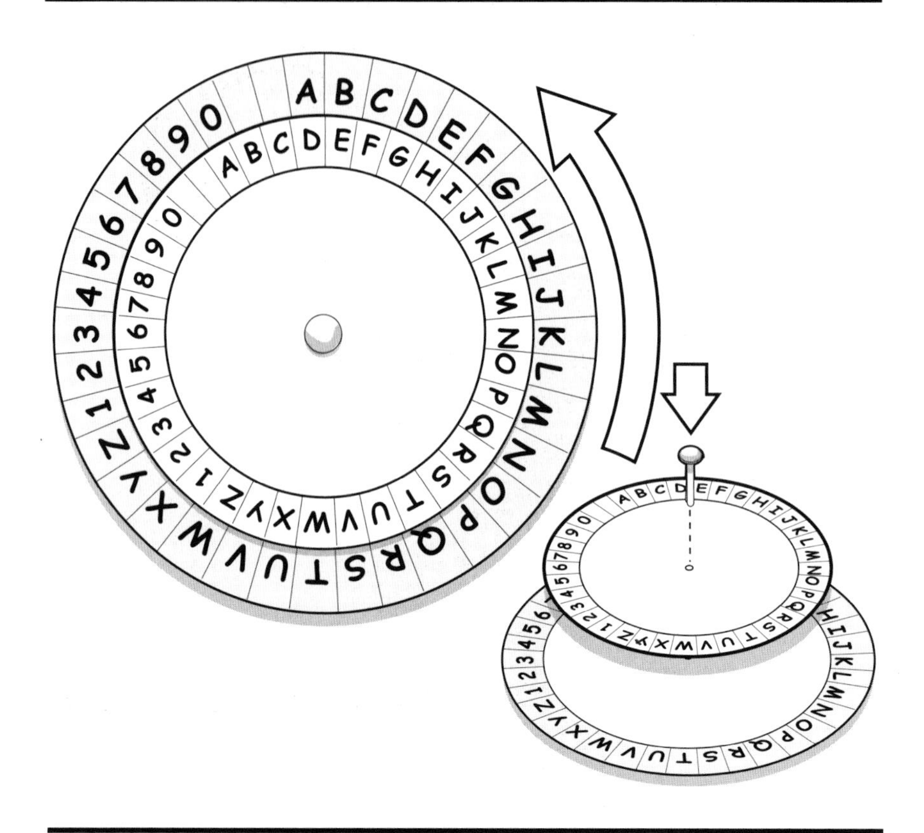

A variation of this code wheel can be constructed from two different diameter circles. The sender of the message lines up a "key" combination (D = A) and then codes the entire note without rotating the wheel for each letter.

The sender must give the receiver a verbal key, which in this case is "D stands for A," so the receiver can decode the message.

KEYHOLE EYE

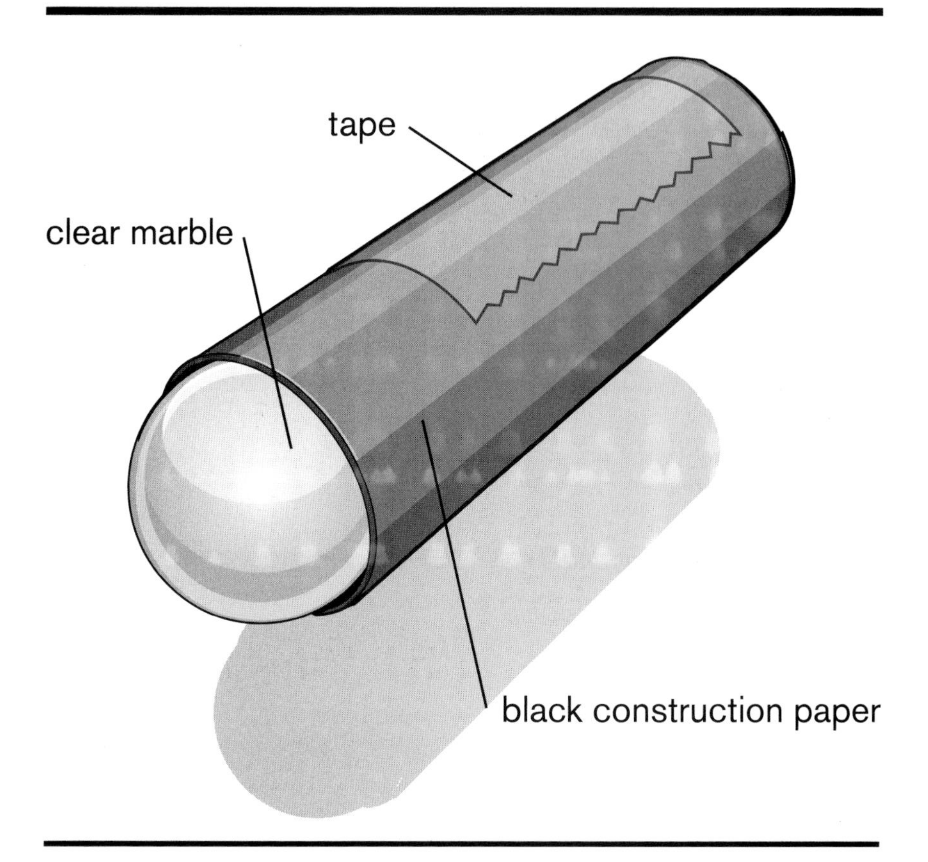

Can you hear whispers behind a closed door but can't pinpoint the source? The Keyhole Eye makes it easy for a spy to observe what's happening, even in hidden rooms. Just slip this pocket-sized gadget into a door gap or access hole and watch closely—the mission might just depend on it! But don't be alarmed that the room is upside down; it's just the marble playing tricks on you.

Supplies

1 sheet of black construction paper
1 small clear marble
Tape (any kind)

Tools

Scissors
Pencil

Step 1

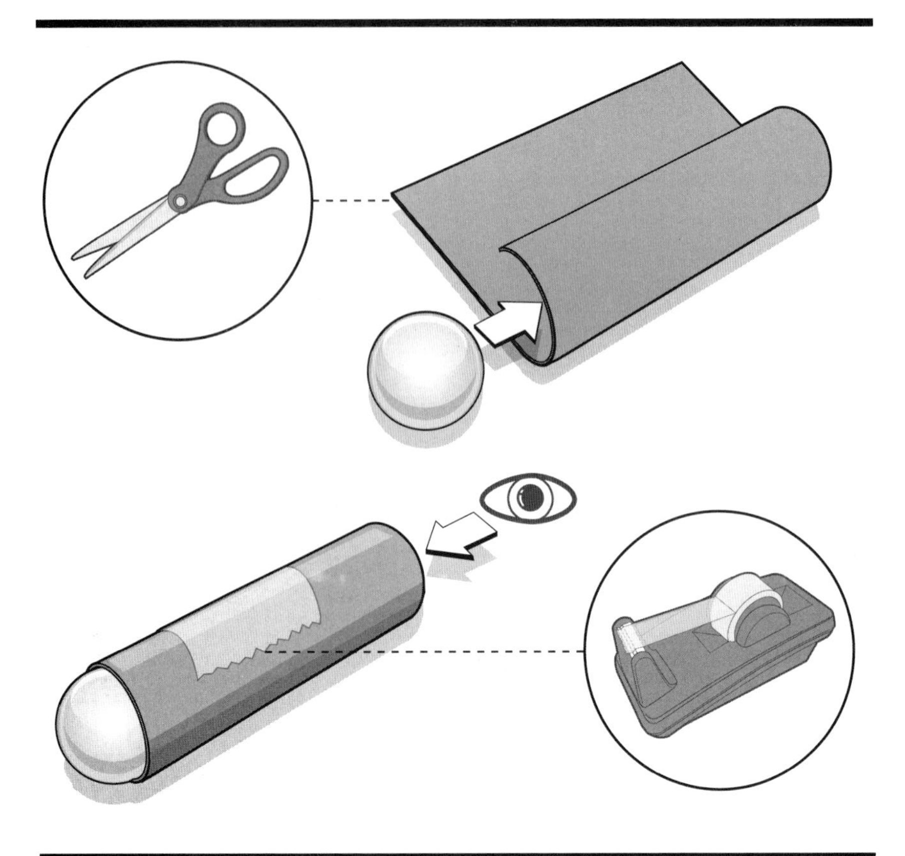

With scissors, cut a piece of black construction paper approximately seven times as long and five times as wide as the clear marble you're using. If black paper is not available, cut a white piece of paper the same size, and then color it with a black marker. The less light that comes through the paper, the better.

When the paper is ready, place the marble at one edge of the construction paper and then tightly roll the black paper around it, creating a tube. Fasten the tube together using tape.

With a pencil, adjust the marble until half of it is sticking out of the constructed tube. Once it's in place, tape around the edge the marble, where it meets the paper, to hold it in firmly. You could also use glue.

To use, slide the marble end through any small door openings and place one eye tight to the tube end. Just remember to remain silent—and never lean against the door, in case your intended target tries to make a quick exit.

6

CONCEALMENT

DECK OF DECEPTION

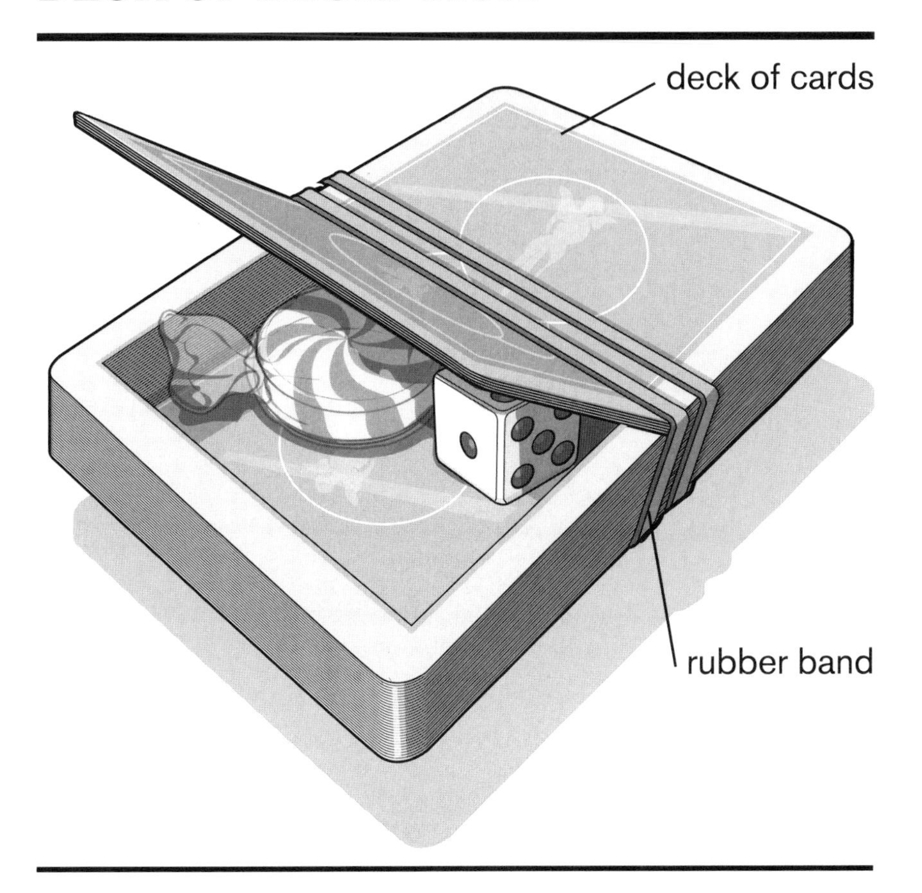

Don't want to get caught transporting unauthorized cargo? The Deck of Deception can help. Its small, hollowed-out cavity is perfect for stashing coordinate chips for top-secret satellites, test tubes filled with clone army DNA, a secret strain of the zombie virus, or candy.

Supplies

1 deck of 52 playing cards
1 piece of scrap paper
1 wide rubber band

Tools

Glue stick
Scissors

Step 1

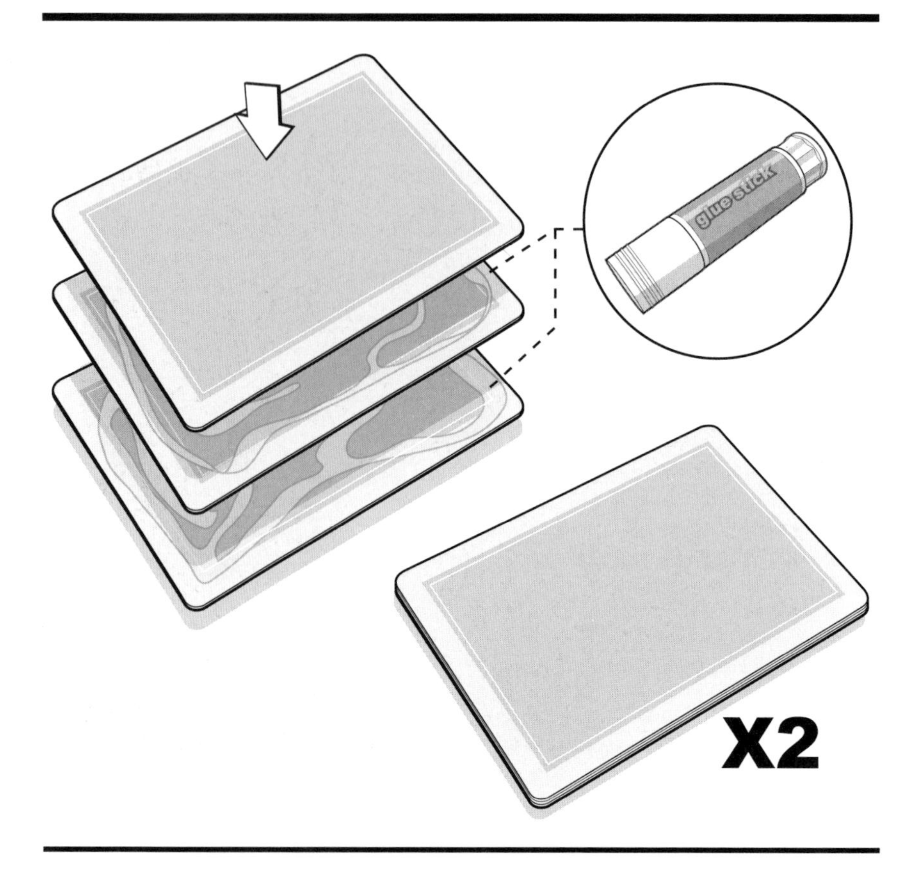

Use a glue stick to attach two sets of three playing cards together, all facing down. These combined sets will make up the bottom and top of your Deck of Deception.

Step 2

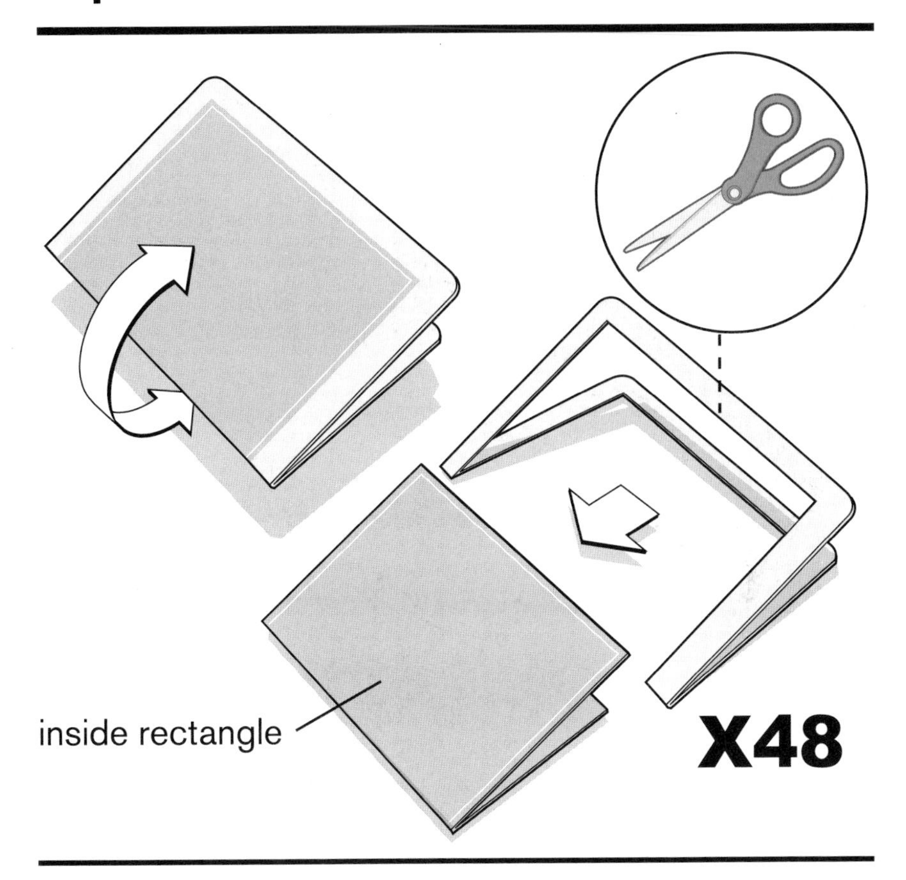

Fold the remaining playing cards (48 if you use the Jokers) in half as shown. Once each card is folded, use scissors to remove a large rectangular hole from inside each card. Use the card's printed design to help make the holes a consistent size. For strength, each frame should be about 1/4 inch wide.

Step 3

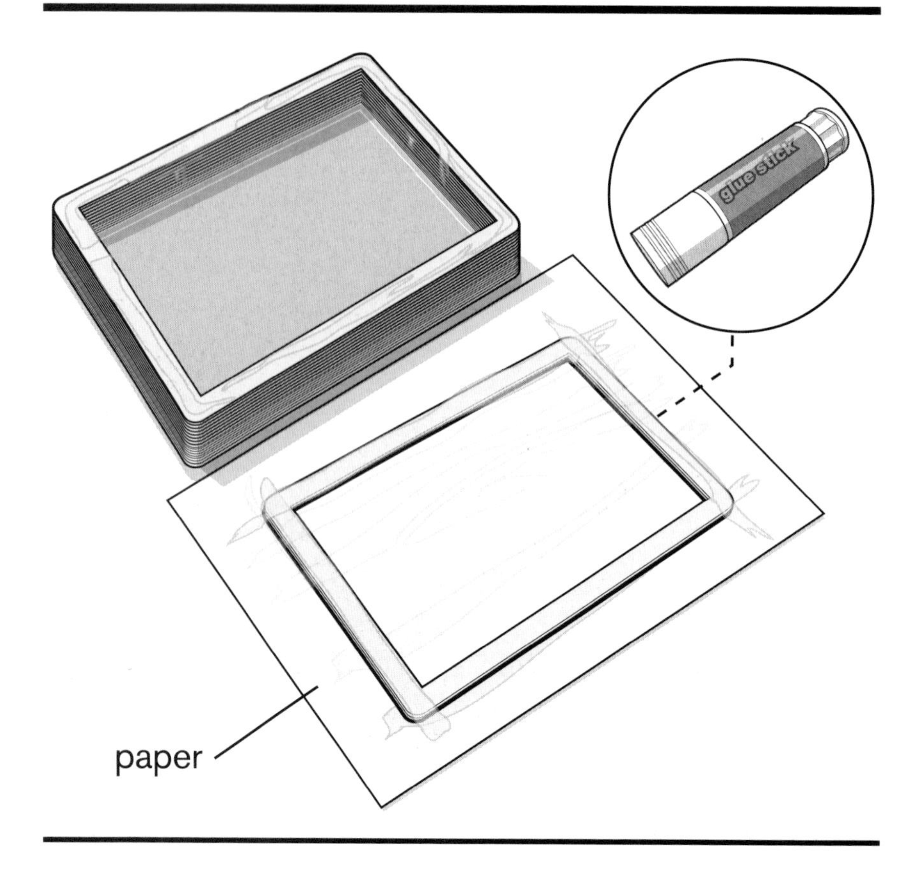

Next, place a sheet of scrap paper on your tabletop surface to keep it clean.

Now it's time to stack the deck! Start with one of the three-card glued sets as the bottom, facedown. Use a glue stick to outline the edge of the top card, then place one framed card on top of the grouping.

Continue to glue all the remaining framed cards, one by one, on top of the base. Make sure each card is aligned with the next, so they appear to be neatly stacked.

Step 4

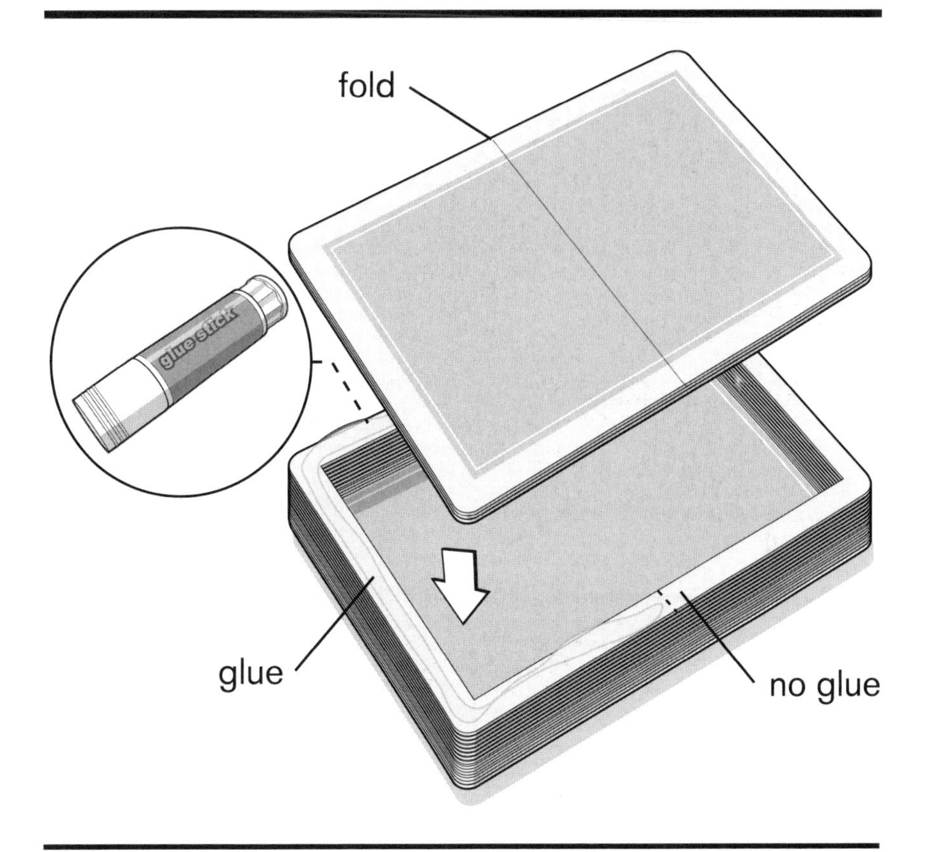

The Deck of Deception is almost complete–it just needs a lid. Glue *half* of the stacked deck, then place the last group of laminated cards on top, facedown.

Step 5

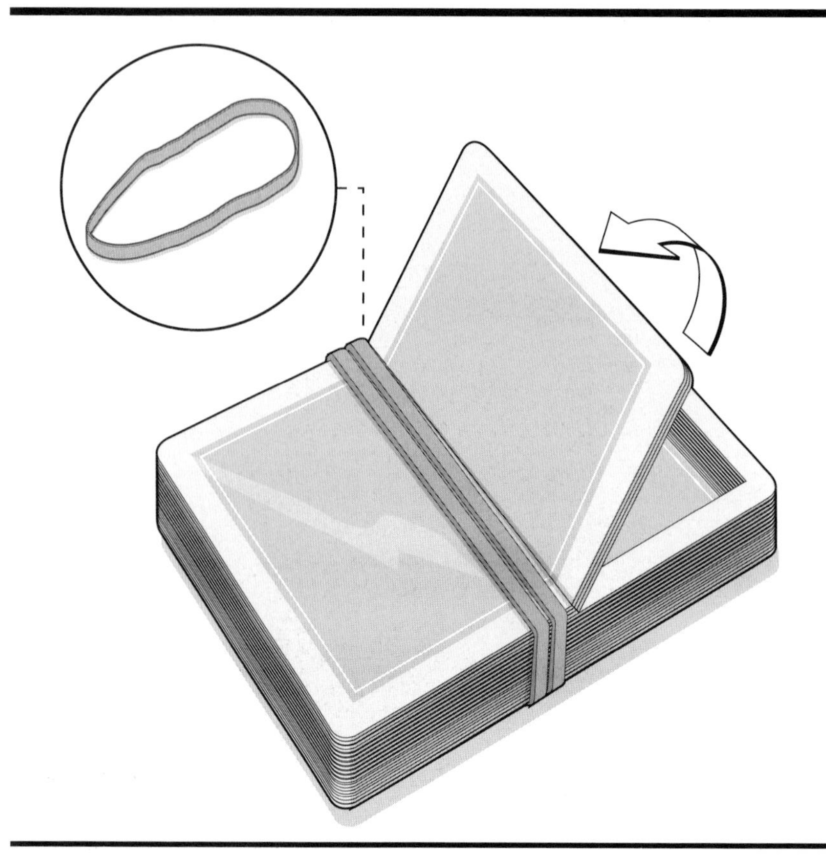

Now tightly wrap a wide rubber band around the card assembly to hold it in place. After the glue has dried, open up the nonglued lid to conceal your prized possessions. Keep the rubber band tightly wrapped around the deck for illusion and deception. (It also hides the folds you made earlier.)

SECRET SOUP SAFE

Afraid your room will be targeted in a shakedown? This Secret Soup Safe will be bypassed during any raid by the opposition. Stuff your valuables in the bottom and then twist it shut. Feebleminded henchmen will be clueless!

Supplies

1 empty soup or vegetable can
1 small glass jar with twist-off cap
Paper towels or newspaper

Tools

Can opener
Hot glue gun
Scissors (optional)

Step 1

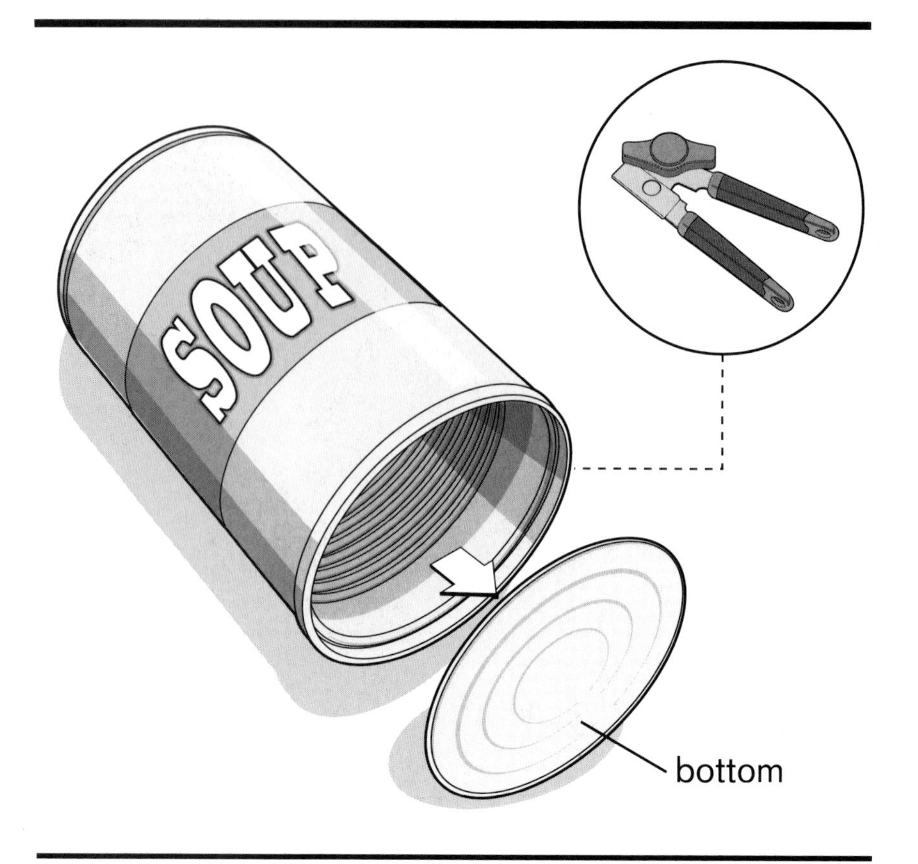

Use a can opener to remove the bottom of a sealed soup or vegetable can. Empty (and eat!) the contents, and wash out the can. Let it dry. Do not discard the bottom.

Step 2

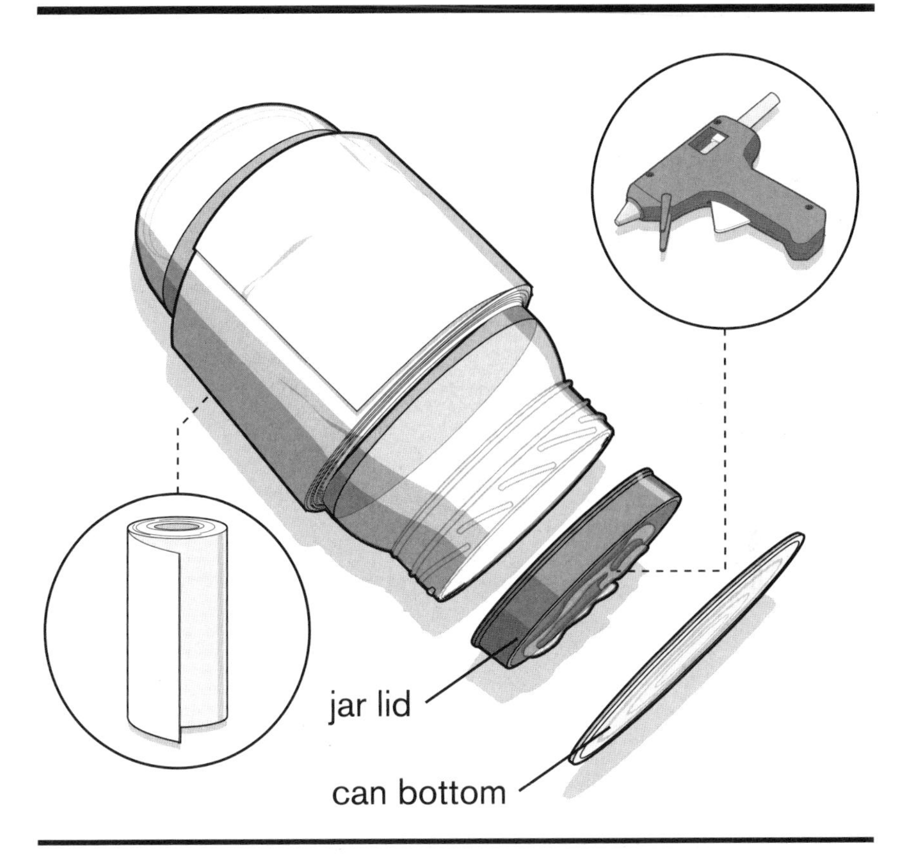

Wash the small glass jar with twist-off cap and let it dry. This jar should fit inside the soup can. Confirm the fit before continuing.

Wrap a paper towel or some newspaper around the glass jar until the jar's outer diameter is equal to the inside diameter of the empty can.

Next, hot glue the can bottom to the center of the jar lid, as shown.

Step 3

Tightly stuff the wrapped jar into the can. Adjust the can so that when you screw the jar lid back onto the jar, the can bottom returns to its original location, flush with the base of the can. If the lid becomes difficult to grasp when screwing it onto the base, use your palm to complete the last few rotations. The Secret Soup Safe is complete. You can load it with valuables and twist it shut.

If the jar rotates when you try to remove the cap, add hot glue to hold it in place.

PLASTIC BOTTLE COMPARTMENT

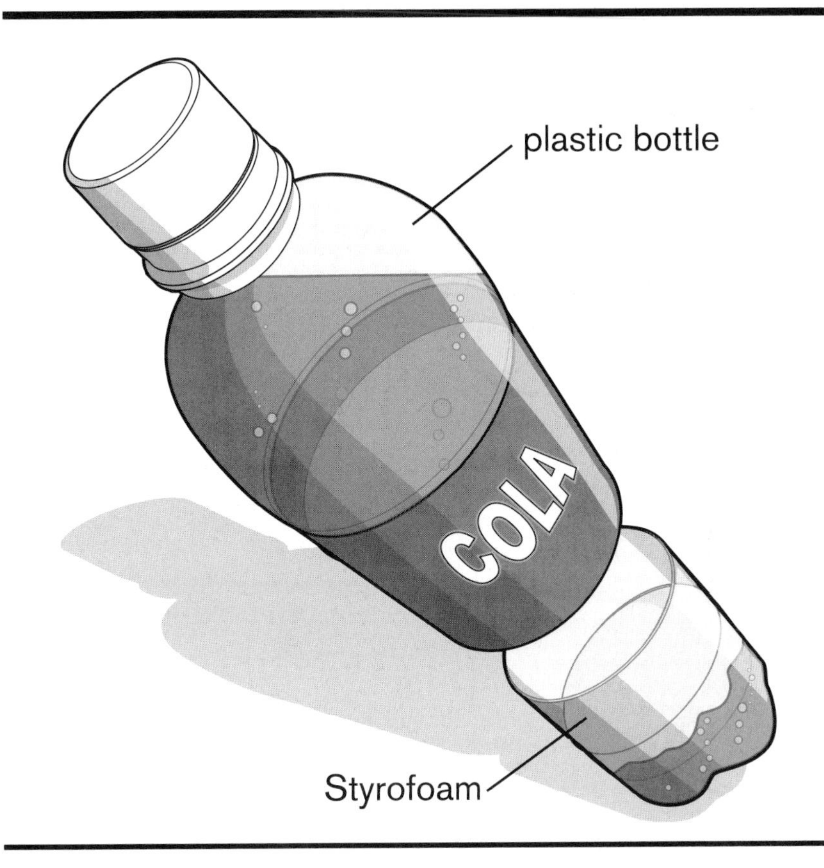

In the field of espionage, things are never what they appear. Secrecy is essential. The Plastic Bottle Compartment is a devious tool used to hide, store, and transport sensitive material. To the curious eye, the bottle appears unmodified and full of liquid, but with a simple twist the bottle reveals a small hidden compartment behind the label area. When constructing this concealment tool, use a beverage that you prefer to drink so as not to attract undue attention.

Supplies

2 plastic bottles, approx. 16.9 fl. oz. (500 mL) each
1 piece of Styrofoam, approx. 1 inch thick

Tools

Large cup
Hobby knife
Scissors
Marker
Hot glue gun

Step 1

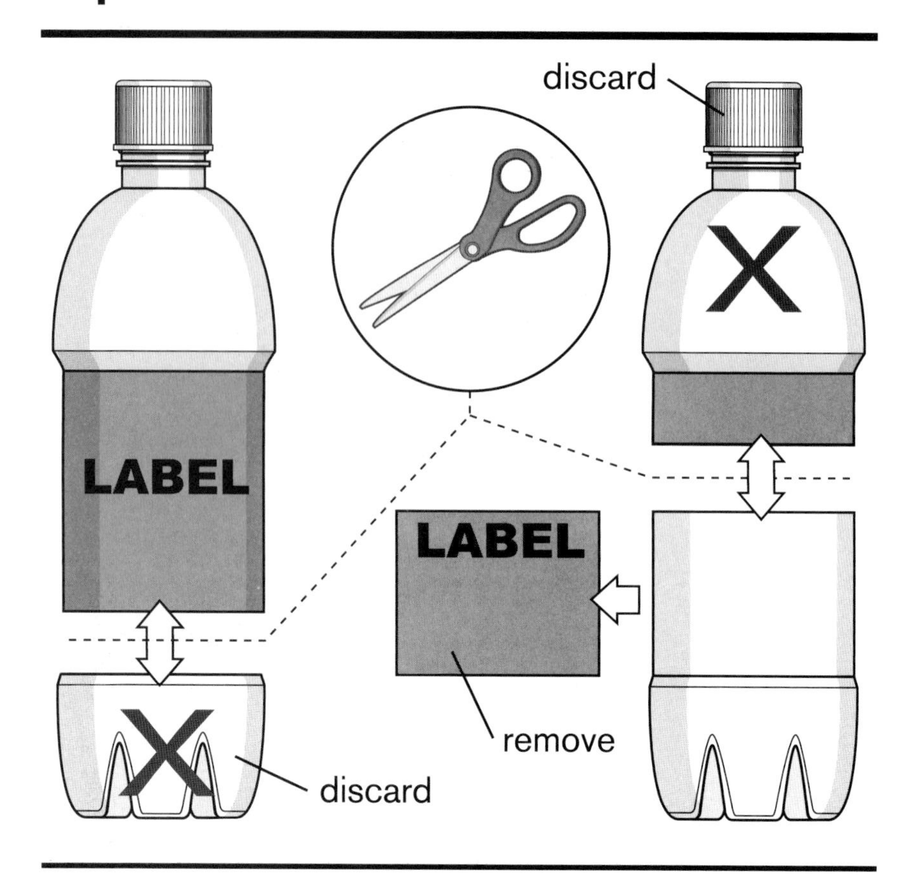

Start with two plastic bottles that are the same size. Empty as much of their liquid as you can into a large cup, then drink what's left. If using a carbonated drink, allow the liquid to go flat before refilling the Plastic Bottle Compartment in steps 3 and 4.

With a hobby knife or scissors, cut the first bottle below the label area as shown on the left side of the illustration. Do not remove the label from this bottle. Discard or recycle its base.

Remove the label from the second bottle. Cut the bottle about 1 inch below where the top of the label rested. Discard the *upper* half of this bottle.

Step 2

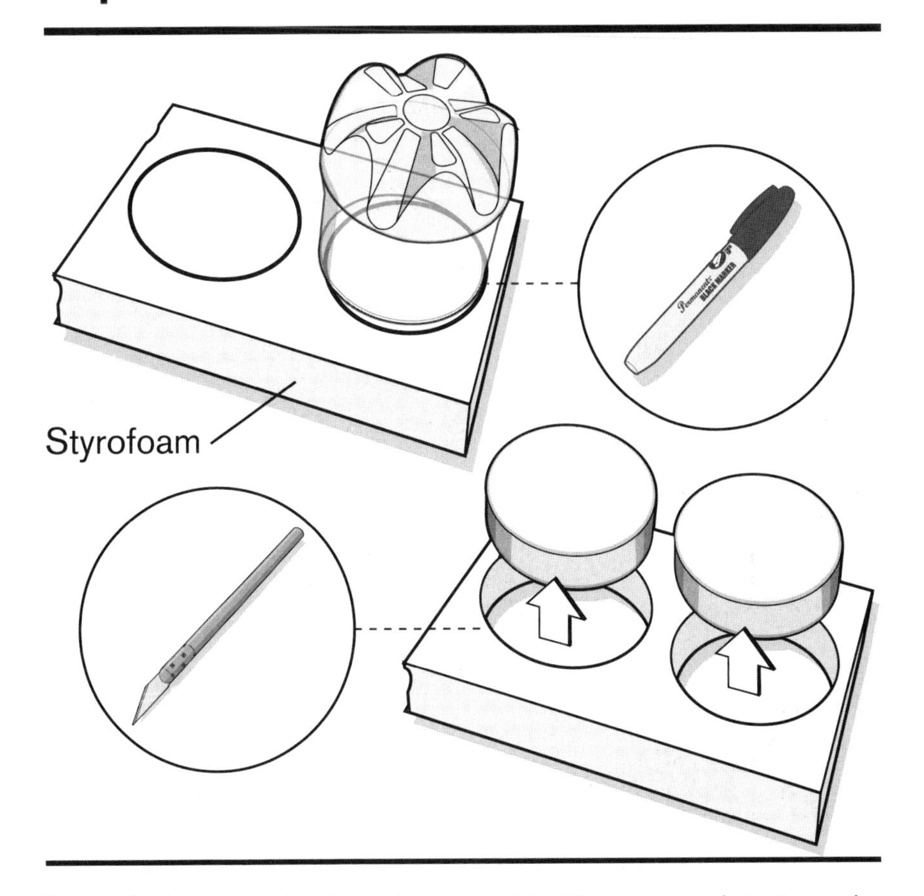

Using the base section from the second bottle as a template, trace the bottle diameter twice onto the Styrofoam. (An old box from an electronic appliance is a good place to find Styrofoam.) Then, cut out the two circles with a hobby knife.

Step 3

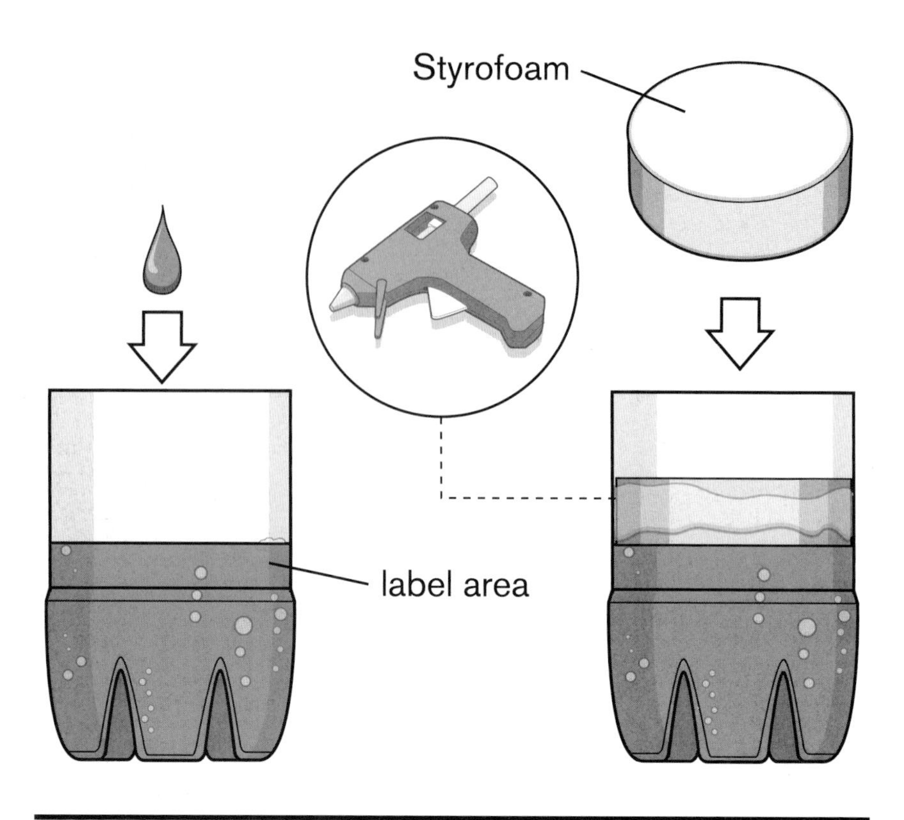

Pour 2 to 3 inches of the original liquid back into the bottom bottle section, as shown. The liquid should stop slightly *above* the removed label area, but no further.

Apply hot glue around the inside of the bottle, just above the liquid, and carefully slide the Styrofoam into the bottle. Push down the Styrofoam until it just touches the liquid. Add an additional ring of hot glue around the top seam to ensure that the seat is watertight.

Step 4

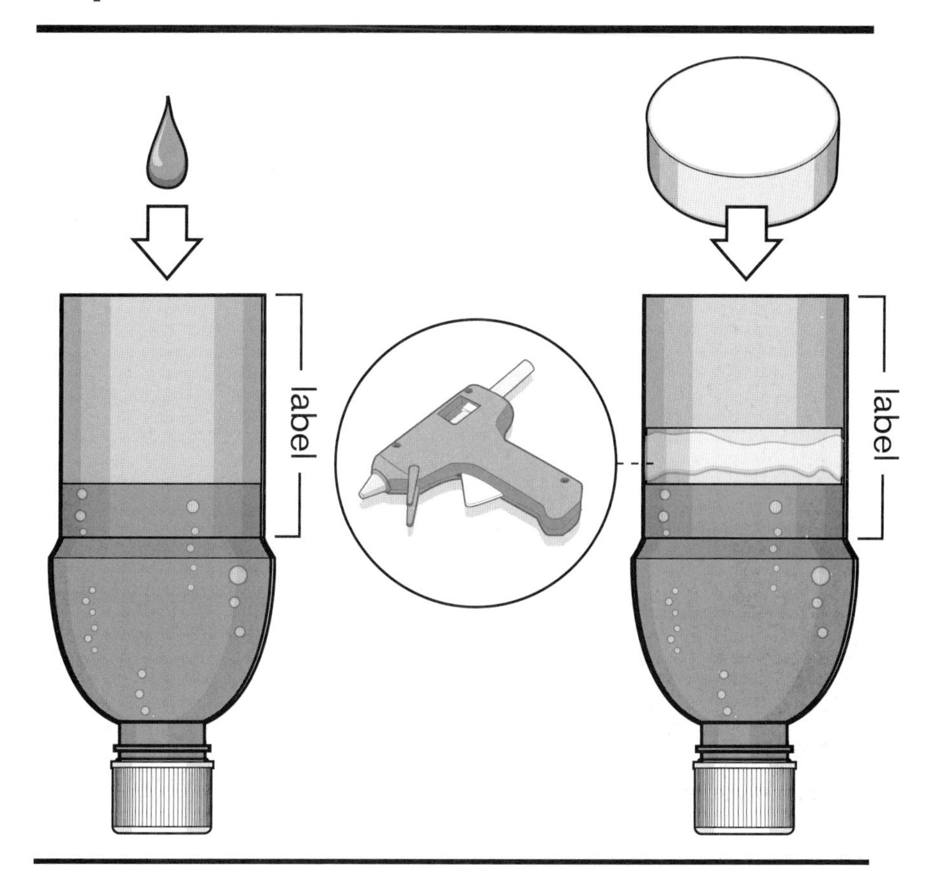

Make sure that the cap on the top section is on tightly, then flip the bottle over, as shown. Pour the original liquid back into the bottle until the liquid has reached the label area (left image).

Carefully apply hot glue around the inside of the bottle just above the liquid, then slide the Styrofoam into place until it just reaches the liquid. Once in place, add more hot glue around the seam to insure that it's watertight. Once the glue cools, turn the bottle over to make sure it doesn't leak, and add additional glue if needed.

You may experience some slight warping of the plastic bottle due to the heat of the glue. This is common, but won't affect the end product.

Step 5

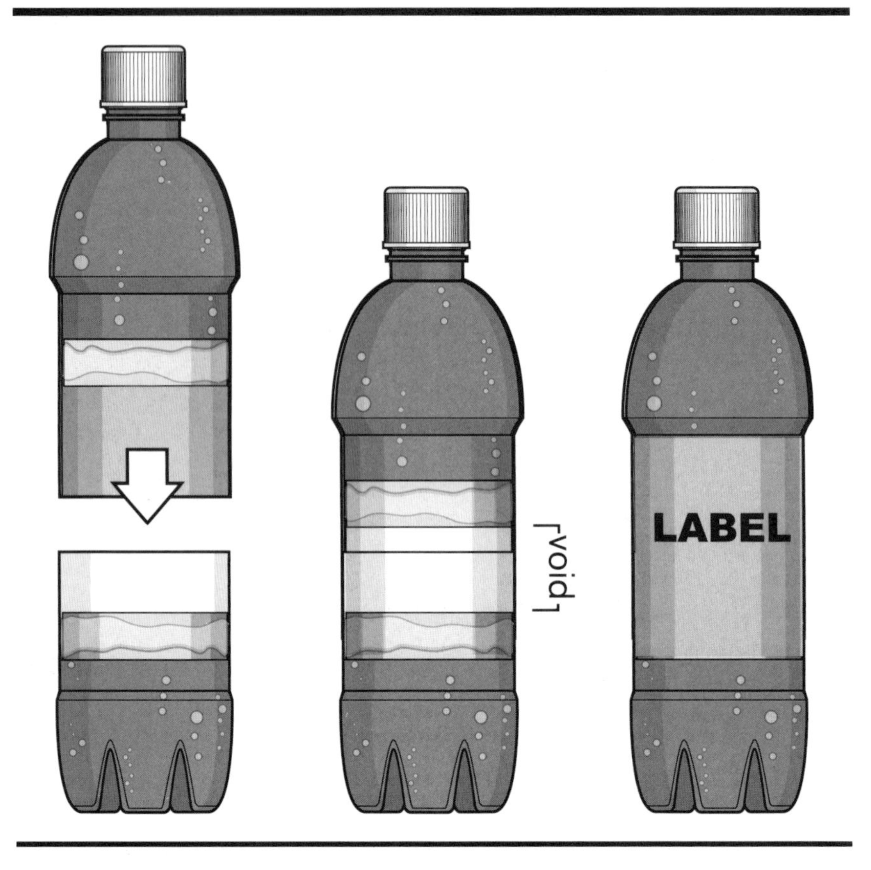

Slide the bottom bottle into the top bottle assembly for a snug fit. When the sections come together they create a small secret compartment, no taller than a few inches. When the sections are connected the bottle looks normal, except for being slightly taller.

Just in case the top hot glue seal fails, it's best not to put anything electronic inside the compartment. To be even safer, seal your secret treasure in a resealable plastic bag and then load it into the bottle compartment.

CEREAL BRIEFCASE

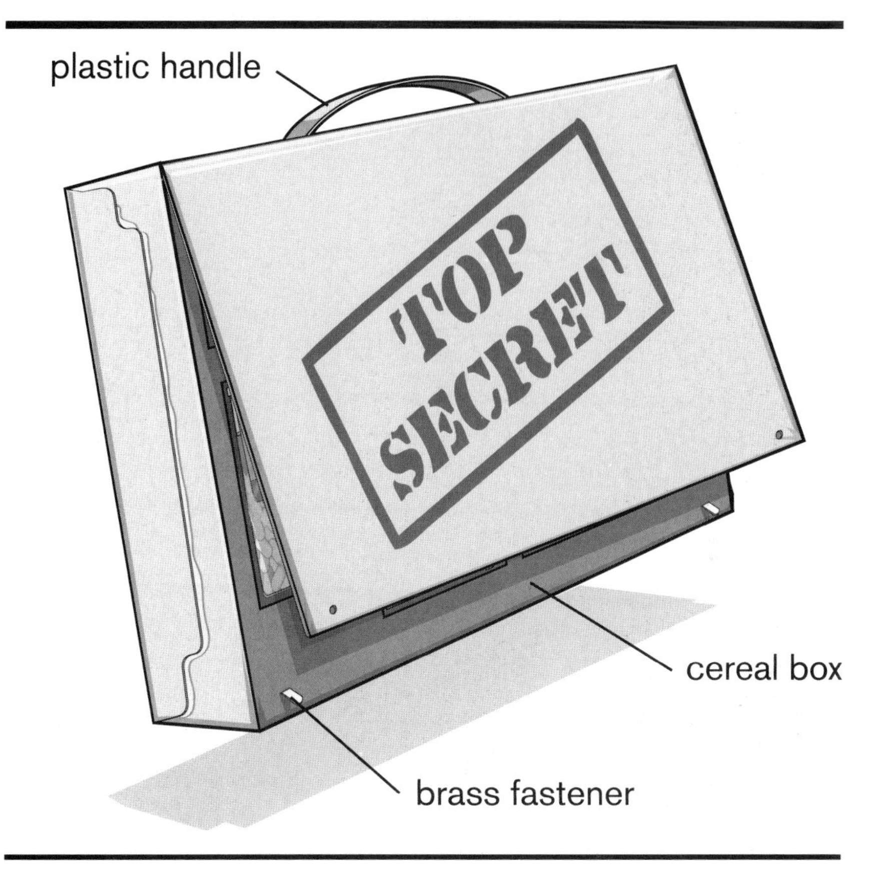

Crafted from a cereal box, the Cereal Briefcase can be customized to store and protect your secret agent arsenal. When the briefcase is closed, your identity and cargo are unknown. Once you're in a safe place, open the briefcase and pick from a library of weapons or gadgets. The briefcase can also be designed to hold extra ammo.

Supplies

1 cereal box
2 sheets of sturdy cardboard, each 1½ feet square
1 plastic handle from a detergent box, shoe box, or computer component box
Duct tape
2 brass fasteners

Tools

Scissors
Hot glue gun
Marker
Hobby knife

Step 1

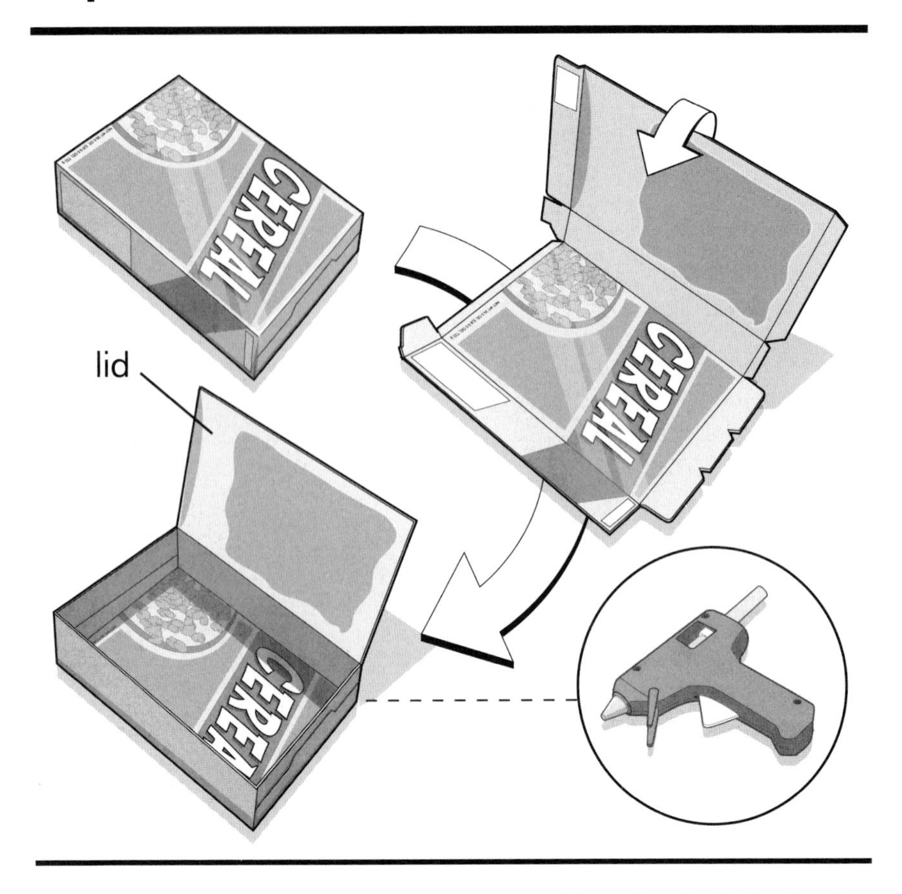

Start with an empty cereal box or cardboard box that is similar in size. Disassemble the box by carefully peeling apart the factory-glued tabs that hold it together. Refold the box, but with the package printing on the inside. Then, using the scissors and hot glue, refasten the box but leave one long side open as a lid. To reassemble the short sides, you may need to cut the flaps off the open side and glue them to rest of the box. Use additional material if needed to fill in any gaps.

Step 2

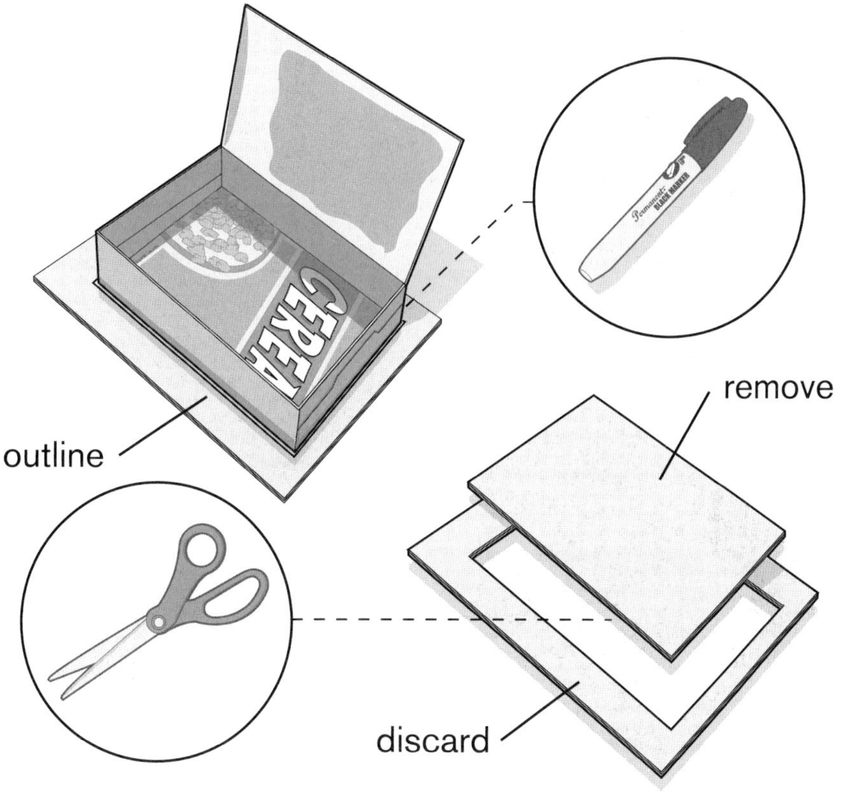

Now place the assembled box on top of a sheet of sturdy cardboard, as shown. Use a marker to outline the box's dimensions onto the cardboard.

Use scissors to cut out the traced rectangle from the cardboard. Save the leftover cardboard for step 4.

Step 3

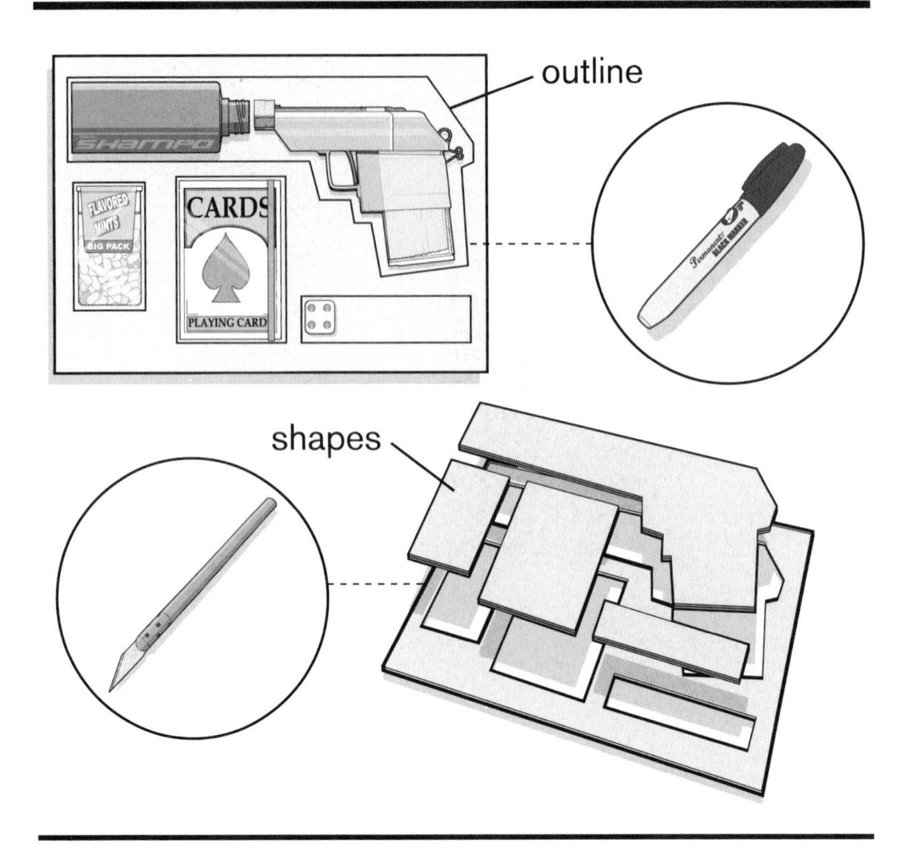

The next step is to create a custom interior that will separate and protect your MiniWeapons and gadgets. Of course, the briefcase will only be able to store items that are smaller than the interior of the box.

Lay out your chosen weapons and gadgets on top of the cardboard rectangle. The illustration above is just an example. Draw a simple outline around each item. Try to minimize the outline detail, making a simple "footprint super" with a minimum number of smooth, straight lines.

Once the outlines are drawn, use a hobby knife to carefully remove the shapes. *Do not* discard these shapes; you will need them in step 6.

Step 4

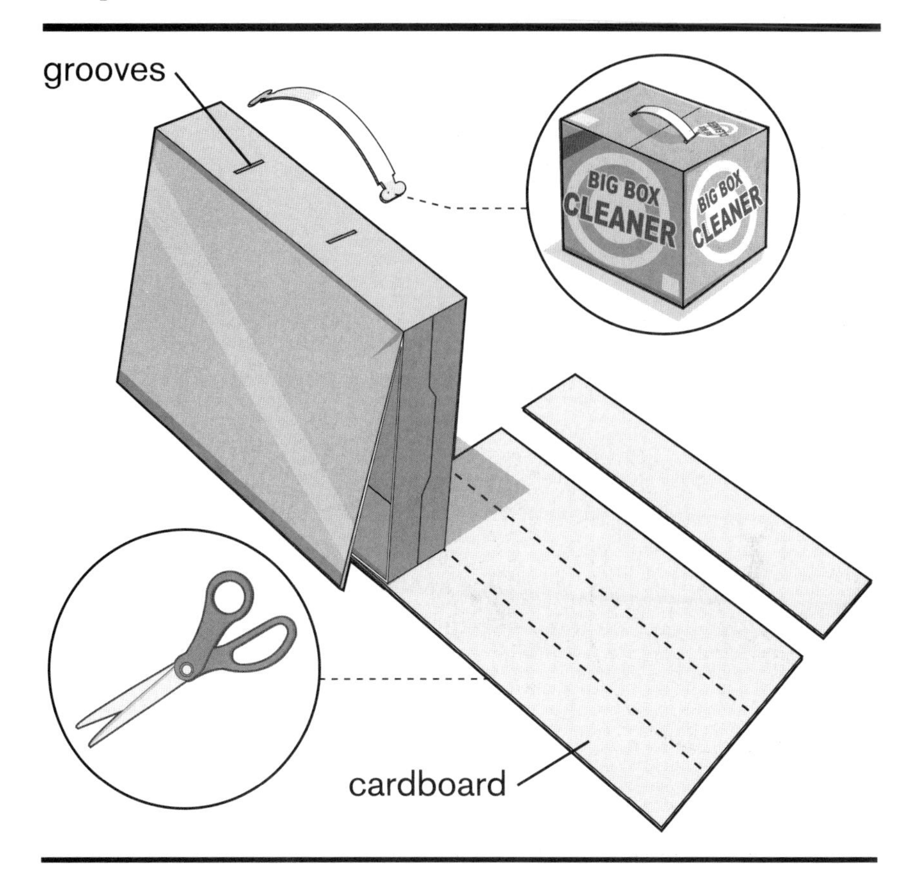

Cut two small slits on the top surface of your briefcase box that will fit the plastic handle. Slide the handle into place, and then tape the handle from the inside of the box to secure it.

Next, place the briefcase box on a new cardboard sheet. Measure out the width of your box and mark it with a pencil, making a strip. Then cut several feet of these strips with a hobby knife or scissors.

Step 5

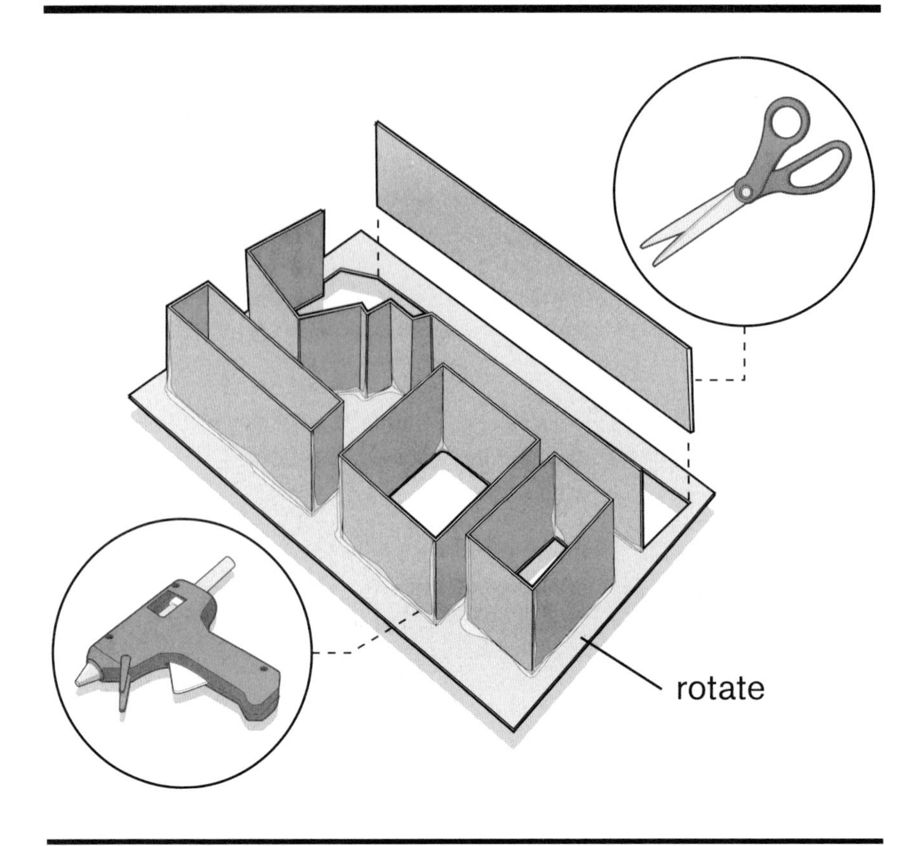

Turn over the cardboard cutout sheet and use the strips of cardboard to outline the *backside* of your weapons pattern, folding, cutting, and gluing the strips for a custom fit. Your craftsmanship doesn't have to be perfect because this is the bottom surface—just make sure you complete each outline. This could take several minutes, so have patience.

Step 6

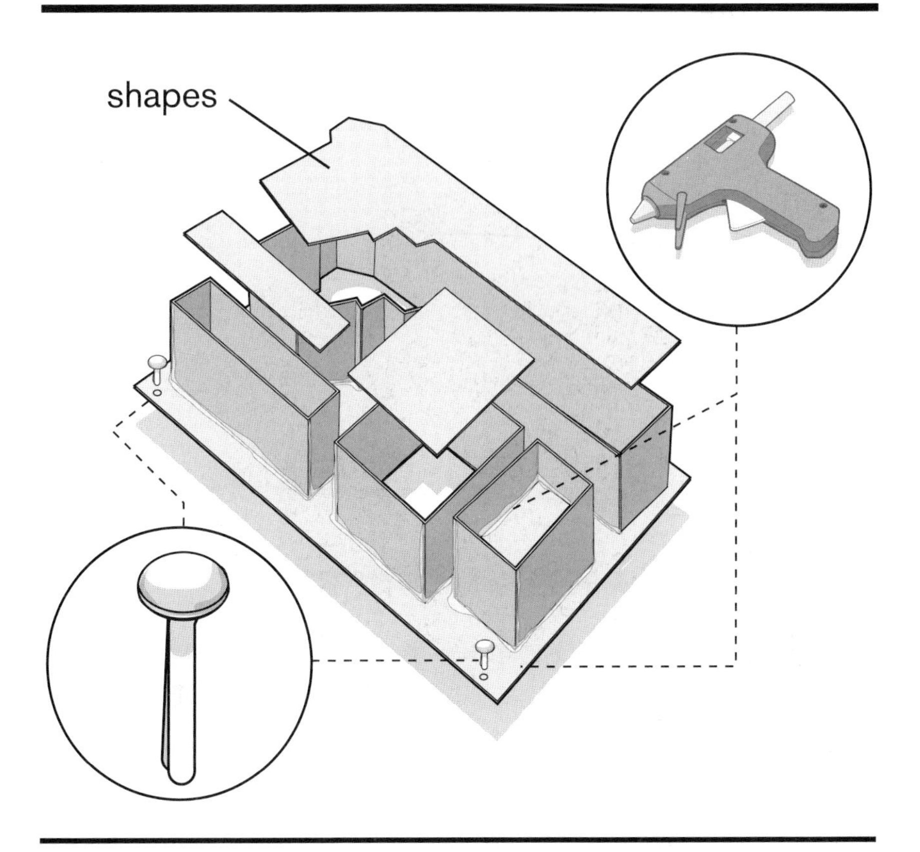

Grab the weapons and gadget outlines you removed in step 3. Slide each shape back into its designated area. Before hot gluing the shapes into place, flip the assembly over so that you can adjust the depth of each shape to accommodate the height of each MiniWeapon. Once you are satisfied, hot glue all the shapes into your assembly.

Add two brass fasteners to the underside of your assembly. Carefully use a hobby knife to cut the holes for both of them at the corners of the cardboard. Once in place, duct tape or hot glue each fastener to the cardboard. These fasteners are a quick and inexpensive way to keep your briefcase lid shut. Velcro strips will also work.

Step 7

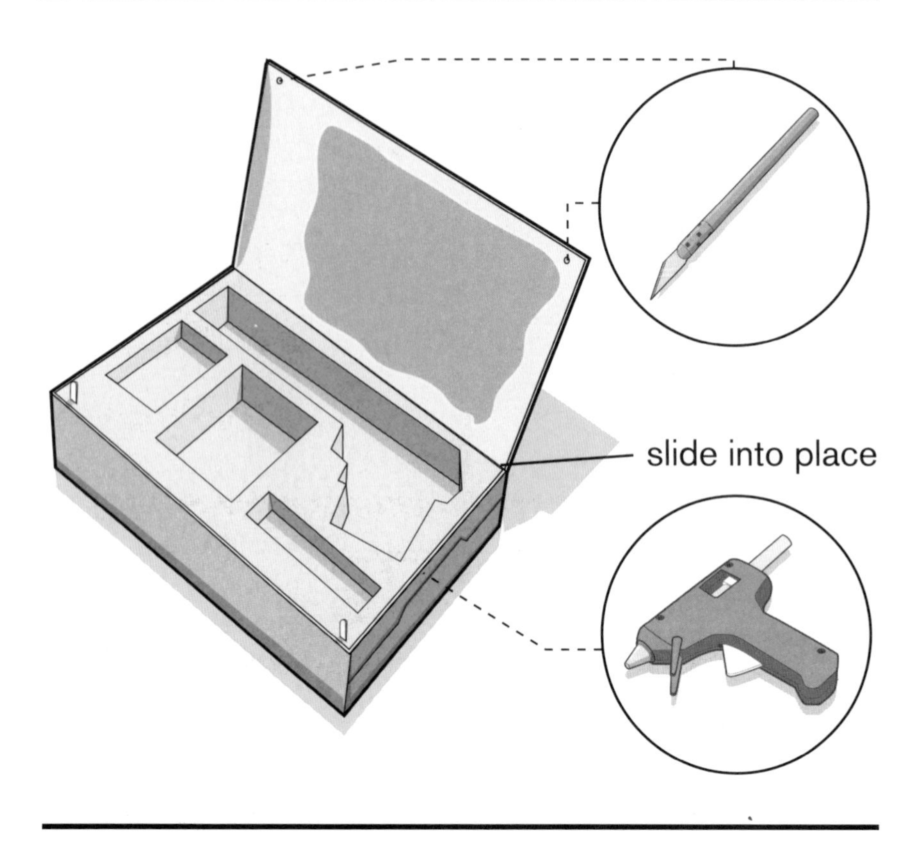

Turn over your assembly and slide it into the briefcase box as shown in the illustration. Test whether the cover can close, and adjust it if needed. Open the lid and glue around the edge of the cardboard to fasten the custom interior to the box's bottom.

Cut two holes in the briefcase cover so that the two brass fasteners can slide though and bend shut.

If you want to go above and beyond the call of duty, customize your Cereal Briefcase by painting it to look like a real briefcase.

7

TARGETS

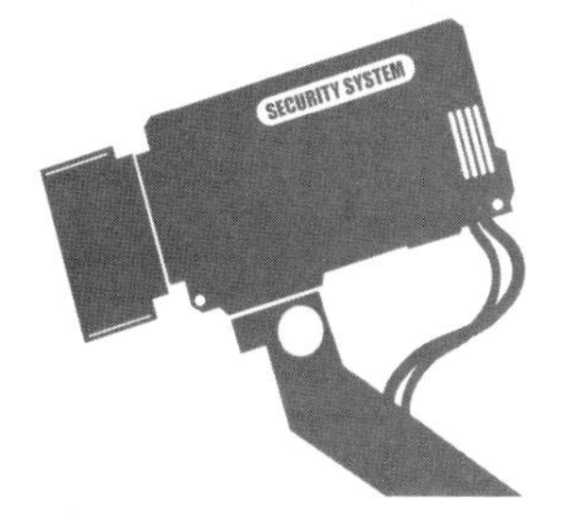

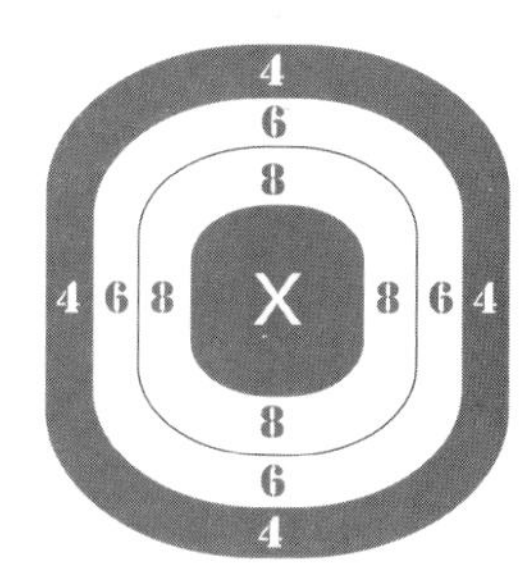

SHARK WITH LASER BEAM

As a spy, you never know what to expect when entering a secret compound. We've heard stories of mutated sea bass, hungry piranhas, and sharks with laser beams.

To earn the coveted *license to kill*, construct this Shark with Laser Beam from a large oatmeal container. Then use it for target practice with your favorite MiniWeapon. Hopefully, this will decrease your chances of becoming a warm meal.

Supplies

1 oatmeal container
1 highlighter cap with clip
2 plastic bottle caps

Tools

Scissors or hobby knife
Hot glue gun
Marker

Step 1

Remove the lid and bottom of a large cylindrical oatmeal container. Use scissors or a hobby knife to cut both ends of the cylinder at an angle to shape the shark's body. Then use that removed material to cut out two triangular pectoral (side) fins and glue both to the cylinder with a hot glue gun. Cut out a dorsal (top) fin from the cylinder as shown, then glue it to the top of the cylinder. The remaining bottom material becomes the caudal (tail) fin. You may need to cut a small slit into the crescent moon–shaped bottom to fit onto the cylinder. Attach the tail fin with hot glue.

With scissors or a hobby knife, cut out the teeth pattern shown above. Slide it into the modified oatmeal cylinder and hot glue it into place. Glue two plastic bottle caps to the top or side of the shark's head and use a marker to draw scary eye details. Finally, clip a highlighter cap (or laser pointer!) onto the dorsal fin to represent a laser beam.

SECURITY CAMERA KNOCKOUT

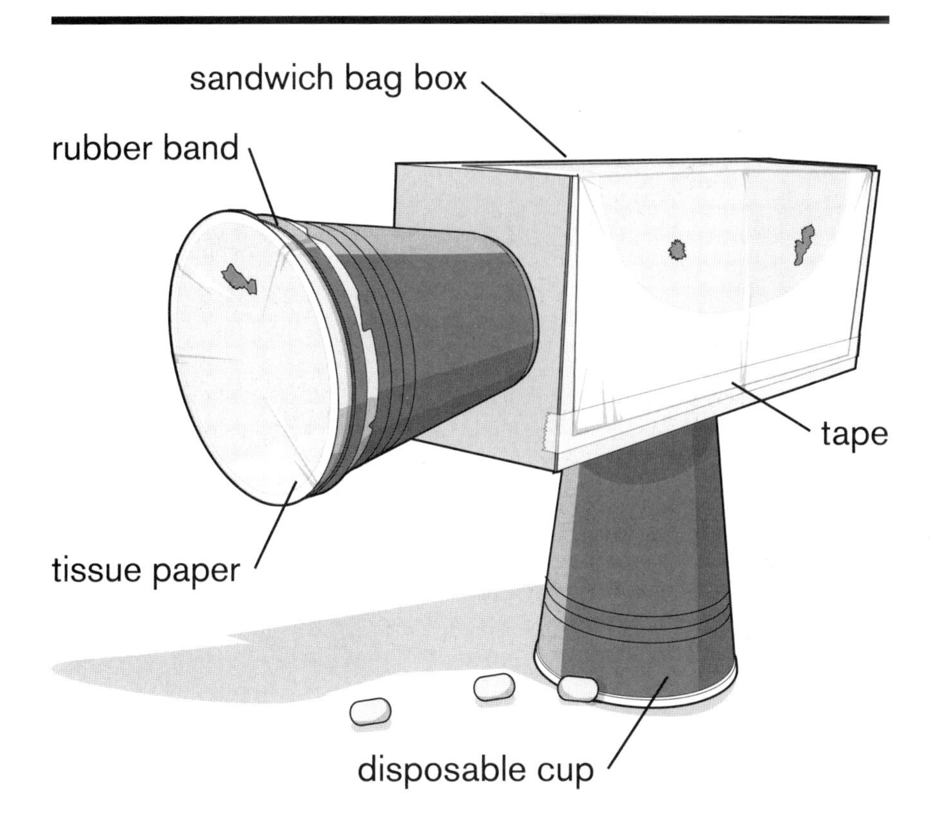

When spying around, watch out for electronic trip wires and security cameras. Trip wires can be avoided, but you'll still need to knock out security cameras.

For practice, build a few of these homemade security cameras and place them around your lair for target practice. They feature a tissue lens and breakaway side panel, perfect for registering a knockout hit from one of your MiniWeapon sidearms. An agent out of sight is an agent out of mind.

Supplies

2 disposable cups
1 sandwich bag box
1+ rubber bands
Tape (any kind)
1+ sheets of bath tissue

Tools

Hot glue gun

Step 1

Glue two disposable cups to the small sandwich bag box, the first cup to the side of the box (for the camera's lens housing) and the second cup beneath the box (for a camera mount). Next, rubber band or tape some tissue over the end of the lens cup. This tissue will make it easy to register a hit. If the box's lid is missing, tape additional tissue over that opening as well; this will be the control panel area.

Tape, binder clip, or balance a few of these homemade cameras around your training facility. Always be aware of your makeshift shooting range; do not place the cameras around or near anything breakable.

KILLER OCTOPUS

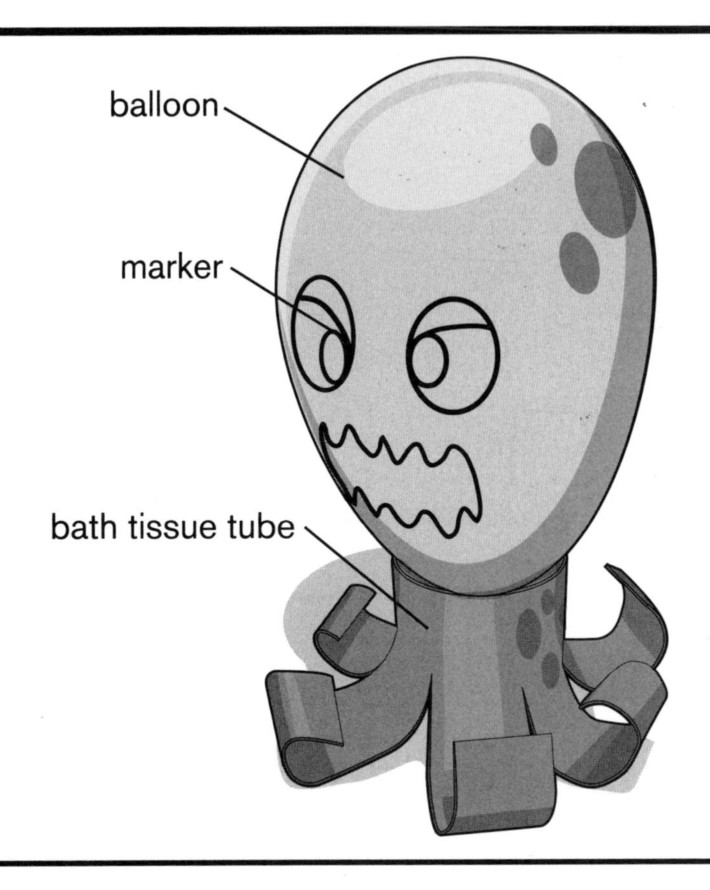

Shhh . . . what was that? *Aaaahhhh!* The radioactive octopuses found us! Mutated to hunt and kill by someone evil, these mysterious octopuses are among the most intelligent of all creatures, land or sea. But during their atomic makeover, someone forgot to add a protective outer shell! With a quick hit from the Q-pick Blowgun, you'll plaster their little latex guts all over the wall! The Killer Octopus is easy to build, using just a balloon and a tissue roll.

Supplies

1 bath tissue roll
1 small balloon
Tape (any kind)

Tools

Scissors
Marker

Step 1

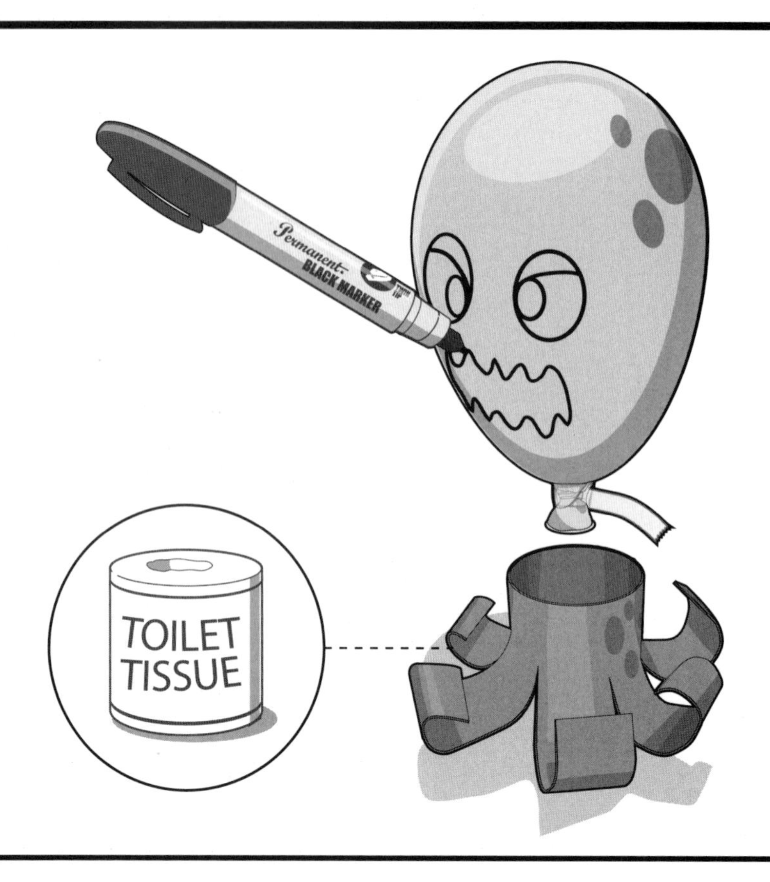

Make several evenly spaced cuts around one end of a bath tissue roll, each about ¾ inch up the tube. Then roll each cut section to form little tentacle ends, as shown in the illustration.

Next, inflate a small balloon and tape the end into the uncut end of the modified tube. Use a marker to draw the octopus's eyes and mouth.

FLIP-DOWN TARGETS

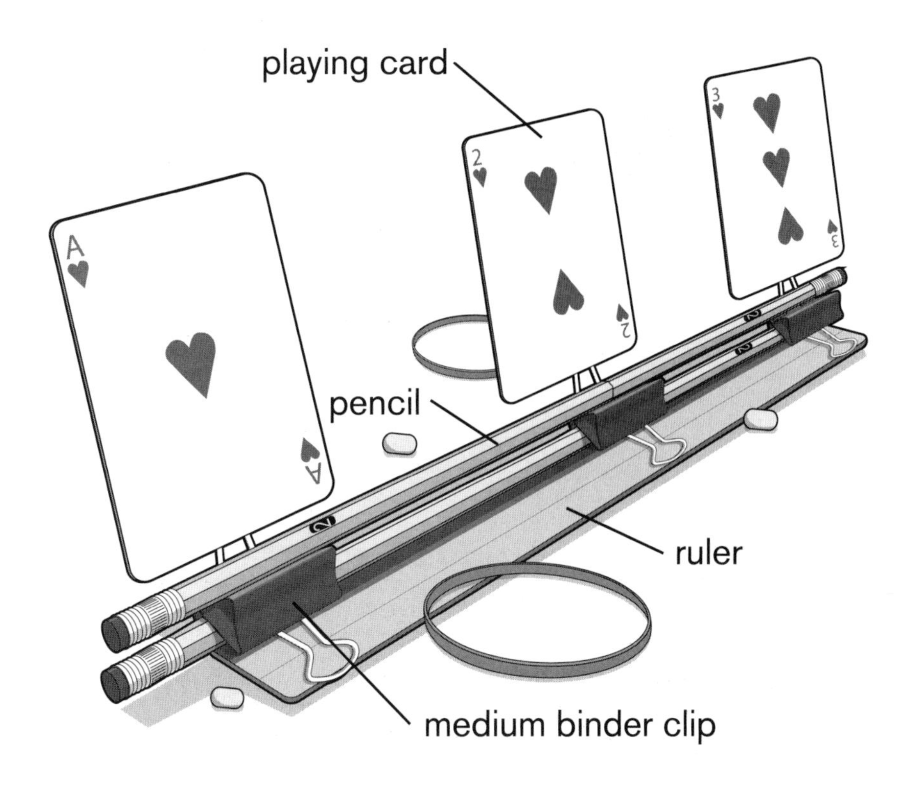

MiniWeapon sharpshooting takes practice, and what better way to unload some ammo than at your very own one-two-three Flip-Down Targets. Fashioned out of a few binder clips, pencils, and playing cards, these targets will help you become a marksman in no time.

Supplies

4 pencils
1 plastic ruler
3 medium binder clips (32 mm)
Tape (any kind)
3 playing cards

Tools

Hot glue gun

Step 1

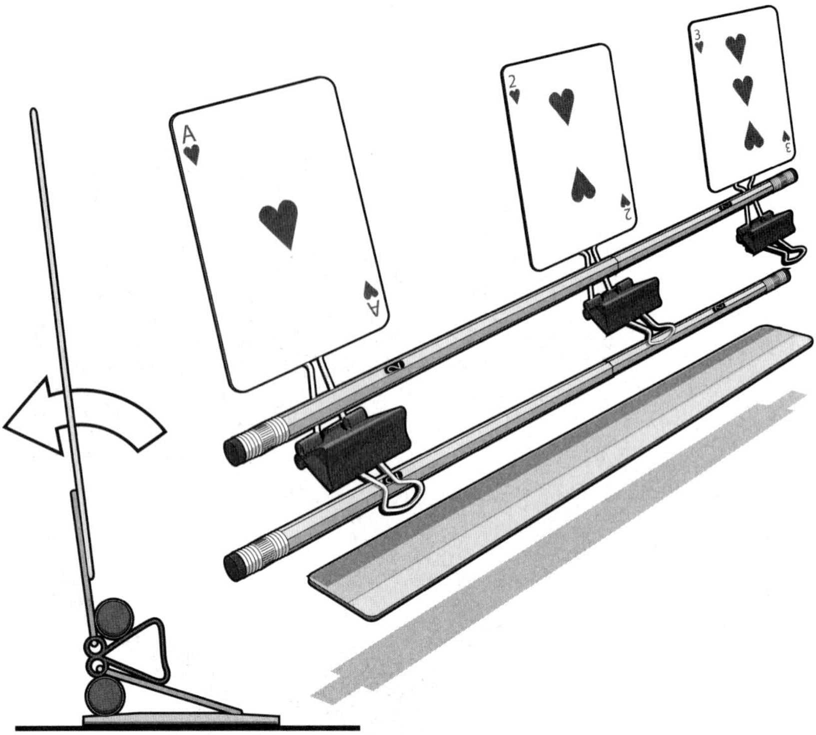

Hot glue two pencils, end to end, to a plastic ruler along one edge. Then hot glue three equally spaced medium binder clips on top of the pencil, with the metal clip handle resting on top of the ruler, as shown. Glue the bottom metal clip handles to the ruler as well.

Now place two more pencils on top of the binder clips, under the metal clip handles. Hot glue only the pencils in place; the clip handles should still rotate. Finally, tape three playing cards to the clip handles as shown.

Your Flip-Down Target system is complete. Never place your target system in front of a backdrop made of breakable materials such as glass, thin wood, or ceramic.

CAMERA TARGET

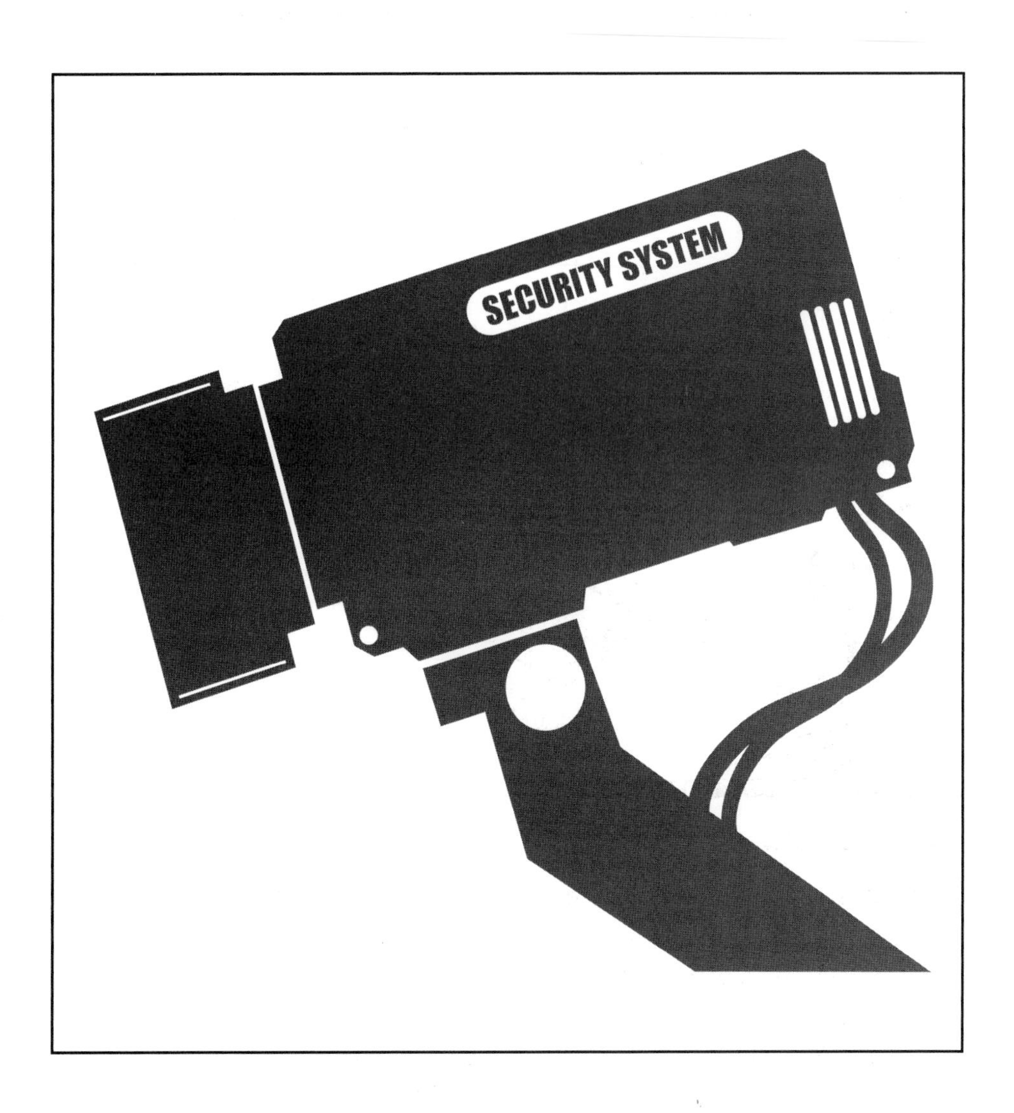

Competitor ______________________________ Date ______

Competitor Signature ______________________________

Use a copy machine to make multiples and enlarge.

EXPLOSIVE BARREL TARGET

Competitor ______________________________ Date ______

Competitor Signature ______________________________

Use a copy machine to make multiples and enlarge.

SIDEARM 10-FT TARGET

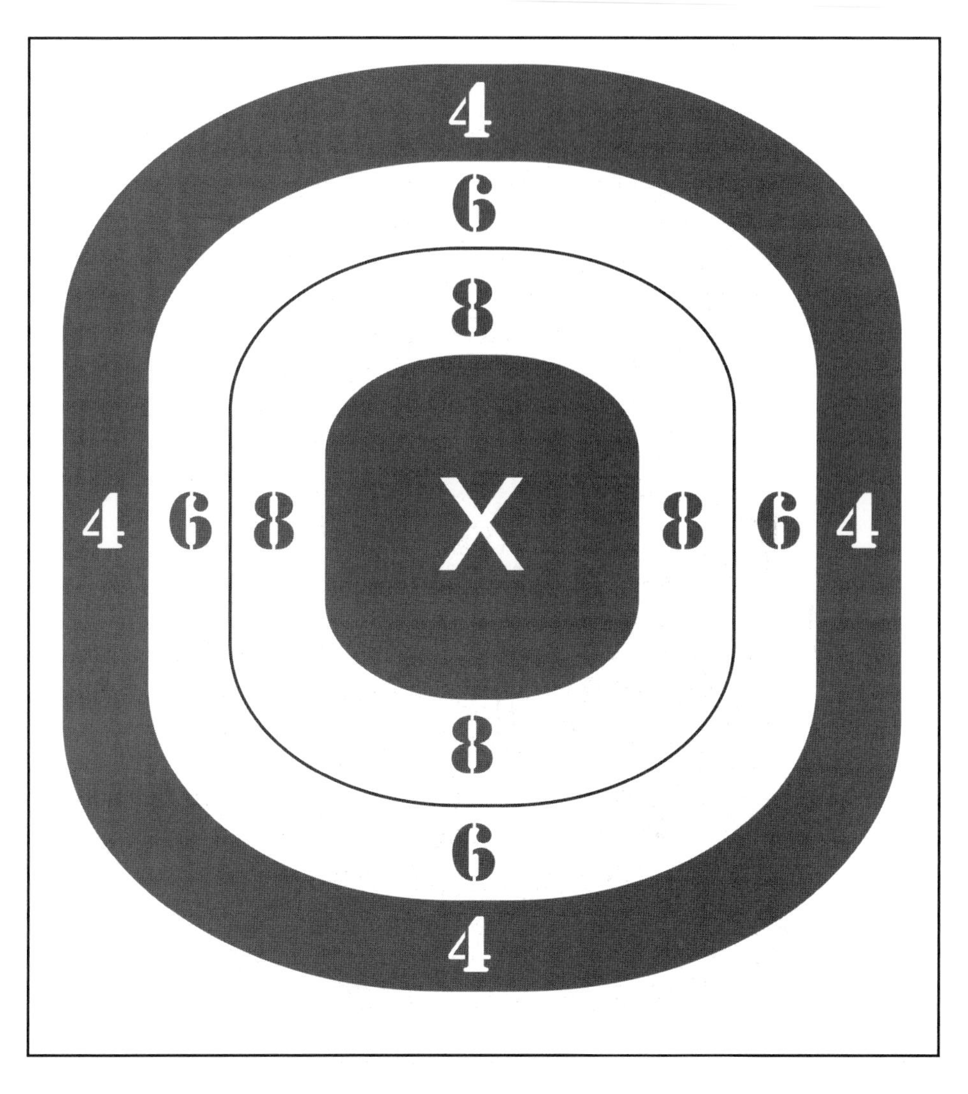

Competitor ______________________________ Date ______

Competitor Signature ________________________________

Use a copy machine to make multiples and enlarge.

For more information and free downloadable targets, please visit:

MINIWEAPONSBOOK.COM

DON'T FORGET TO JOIN THE MINIWEAPONS ARMY ON FACEBOOK:

MiniWeapons of Mass Destruction: Homemade Weapons Page

ALSO FROM CHICAGO REVIEW PRESS

So Now You're a Zombie

A Handbook for the Newly Undead

John Austin

978-1-56976-342-1
$14.95 (CAN $16.95)

Zombies know that being undead can be disorienting. Your arms and other appendages tend to rot and fall off. It's difficult to communicate with a vocabulary limited to moans and gurgles. And that smell! (Yes, it's *you*.) But most of all, you must constantly find and ingest human brains. *Braaaains!!!*

What's a reanimated corpse to do?

As the first handbook written specifically for the undead, *So Now You're a Zombie* explains how your new, putrid body works and what you need to survive in this zombiphobic world. Dozens of helpful diagrams outline attack strategies to secure your human prey, such as the Ghoul Reach, the Flanking Zeds, the Bite Hold, and the Aerial Fall. You'll learn how to successfully extract the living from boarded-up farmhouses and broken-down vehicles. Zombiologist John Austin even explores the upside of being a zombie. Gone are the burdens of employment, taxes, social networks, and basic hygiene, allowing you to focus on the simple necessities: the juicy gray matter found in the skulls of the living.

Practical Pyromaniac, The

Build Fire Tornadoes, One-Candlepower Engines, Great Balls of Fire, and More Incendiary Devices

William Gurstelle

978-1-56976-710-8
$16.95 (CAN $18.95)

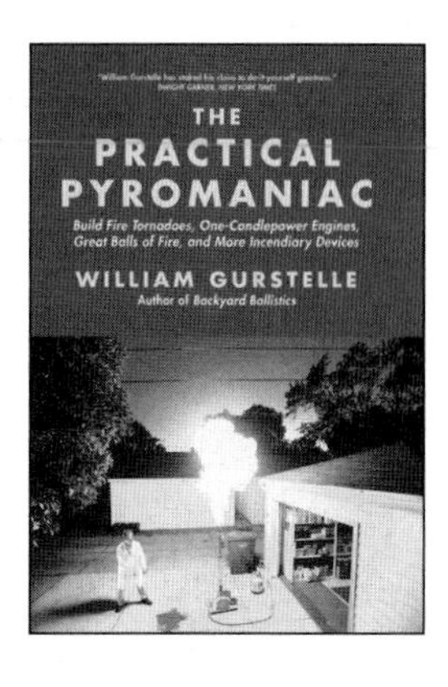

"What a fun, totally engrossing book! Gurstelle's projects—everything from a tiny single-candle engine to a flamethrower—are both easy to build and hard to resist. . . . Think of *The Practical Pyromaniac* as a cookbook for the budding scientist in each of us." –James Meigs, editor in chief of *Popular Mechanics*

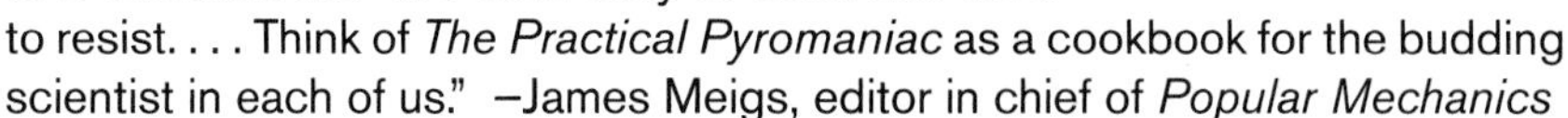

The Practical Pyromaniac combines science, history, and do-it-yourself pyrotechnics to explain humankind's most useful and paradoxical tool: fire. William Gurstelle, frequent contributor to *Popular Mechanics* and *Make* magazine, presents dozens of projects with instructions, diagrams, photos, and links to video demonstrations that enable people of all ages (including young enthusiasts with proper supervision) to explore and safely play with fire.

Backyard Ballistics

Build Potato Cannons, Paper Match Rockets, Cincinnati Fire Kites, Tennis Ball Mortars, and More Dynamite Devices

William Gurstelle

978-1-55652-375-5
$16.95 (CAN $18.95)

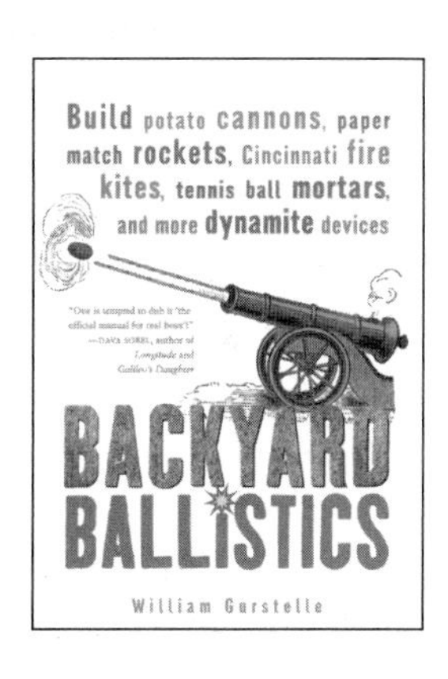

"If you want to make a potato souffle, pick up a book by Julia Child. If you want to decorate your holiday cards with hand-cut potato stamps, look to a Martha Stewart manual. If, however, you'd like to launch a potato in a blazing fireball of combusting hairspray from a PVC pipe, your best source is *Backyard Ballistics*." –*Time Out New York*

Ordinary folks can construct 13 awesome ballistic devices in their garage or basement workshops using inexpensive household or hardware-store materials and this step-by-step guide. Clear instructions, diagrams, and photographs show how to build projects ranging from the simple—a match-powered rocket—to the more complex—a scale-model tabletop catapult—to the offbeat—a tennis-ball cannon.

The Art of the Catapult

Build Greek Ballistae, Roman Onagers, English Trebuchets, and More Ancient Artillery

William Gurstelle

978-1-55652-526-1
$16.95 (CAN $18.95)

"This book is a hoot . . . the modern version of *Fun for Boys* and *Harper's Electricity for Boys*."
—*Natural History*

Whether playing at defending their own castle or simply chucking pumpkins over a fence, wannabe marauders and tinkerers will become fast acquainted with Ludgar, the War Wolf, Ill Neighbor, Cabulus, and the Wild Donkey—ancient artillery devices known commonly as catapults. Instructions and diagrams illustrate how to build seven authentic, working model catapults, including an early Greek ballista, a Roman onager, and the apex of catapult technology, the English trebuchet.

Soda-Pop Rockets

20 Sensational Rockets to Make from Plastic Bottles

Paul Jarvis

978-1-55652-960-3
$16.95 (CAN $18.95)

Anyone can recycle a plastic bottle by tossing it into a bin, but it takes a bit of skill to propel it into a bin from 500 feet away. This fun guide features 20 different easy-to-launch rockets that can be built from discarded plastic drink bottles. After learning how to construct and launch a basic model, you'll find new ways to modify and improve your designs. Clear, step-by-step instructions with full-color illustrations accompany each project, along with photographs of the author firing his creations into the sky.

The Paper Boomerang Book

Build Them, Throw Them, and Get Them to Return Every Time

Mark Latno

978-1-56976-282-0
$12.95 (CAN $13.95)

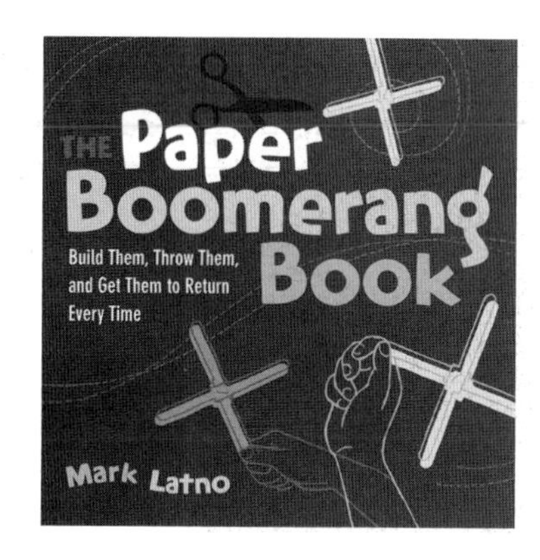

The Paper Boomerang Book is the first-of-its-kind guide to this fascinating toy. Boomerang expert Mark Latno will tell you how to build, perfect, and troubleshoot your own model. Once you've mastered the basic throw, return, and catch, it's on to more impressive tricks–the Over-the-Shoulder Throw, the Boomerang Juggle, the Under-the-Leg Catch, and the dreaded Double-Handed, Backward, Double-Boomerang Throw. And best of all, you don't have to wait for a clear, sunny day to test your flyers–they can be flown indoors in almost any sized room, rain or shine.

Mondo Magnets

40 Attractive (and Repulsive) Devices & Demonstrations

Fred Jeffers

978-1-55652-630-5
$16.95 (CAN $22.95)

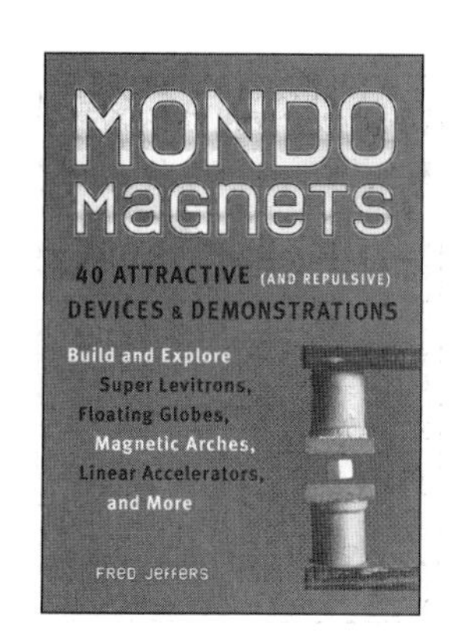

Surprising and seemingly impossible effects result from the 40 experiments included in this fascinating guide to the wondrous world of magnetism. Each experiment–such as using a common refrigerator magnet to create a three-dimensional image or floating a magnet and carbon sheet in midair–is outlined with step-by-step instructions and diagrams. Even the most experienced at-home tinkerer will find dozens of new tricks in this amazing collection.

Gonzo Gizmos

Projects & Devices to Channel Your Inner Geek

Simon Field

978-1-55652-520-9
$16.95 (CAN $18.95)

This book for workbench warriors and grown-up geeks features step-by-step instructions for building more than 30 fascinating devices. Detailed illustrations and diagrams explain how to construct a simple radio with a soldering iron, a few basic circuits, and three shiny pennies; how to create a rotary steam engine in just 15 minutes with a candle, a soda can, and a length of copper tubing; and how to use optics to roast a hot dog, using just a flexible plastic mirror, a wooden box, a little algebra, and a sunny day. Also included are experiments most science teachers probably never demonstrated, such as magnets that levitate in mid-air, metals that melt in hot water, a Van de Graaff generator made from a pair of empty soda cans, and lasers that transmit radio signals.

Return of Gonzo Gizmos

More Projects & Devices to Channel Your Inner Geek

Simon Field

978-1-55652-610-7
$16.95 (CAN $22.95)

This fresh collection of more than 20 science projects—from hydrogen fuel cells to computer-controlled radio transmitters—is perfect for the tireless tinkerer. Its innovative activities include taking detailed plant cell photographs through a microscope using a disposable camera; building a rocket engine out of aluminum foil, paper clips, and kitchen matches; and constructing a geodesic dome out of gumdrops and barbecue skewers. Most of the devices can be built using common household products or components available at hardware or electronic stores, and each experiment contains illustrated step-by-step instructions with photographs and diagrams that make construction easy.

Haywired

Pointless (Yet Awesome) Projects for the Electronically Inclined

Mike Rigsby

978-1-55652-779-1
$16.95 (CAN $18.95)

"A great first book for the budding inventor or engineer." –Eric Wilhelm, cofounding partner of Squid Labs, CEO of Instructables

Written for budding electronics hobbyists, *Haywired* proves that science can inspire odd contraptions. Create a Mona Lisa that smiles even wider when you approach it. Learn how to build and record a talking alarm or craft your own talking greeting card. Construct a no-battery electric car toy that uses a super capacitor, or a flashlight that can be charged in minutes then shine for 24 hours. Each project is described in step-by-step detail with photographs and circuit diagrams, and helpful hints are provided on soldering, wire wrapping, and multimeter use.

Unscrewed

Salvage and Reuse Motors, Gears, Switches, and More from Your Old Electronics

Ed Sobey

978-1-56976-604-0
$16.95 (CAN $18.95)

Unscrewed is the perfect resource for all UIYers–Undo It Yourselfers–looking to salvage hidden treasures or repurpose old junk. Author Ed Sobey will show you how to safely disassemble more than 50 devices, from laser printers to VCRs to radio-controlled cars. Each deconstruction project includes a "treasure cache" of the components to be found, a required tools list, and step-by-step instructions, with photos, on how to extract the working components. It also includes suggestions on how to repurpose your electronic finds. Fight the mindset of planned obsolescence–there's technological gold in that there junk!

The Way Toys Work

The Science Behind the Magic 8 Ball, Etch A Sketch, Boomerang, and More

Ed Sobey and Woody Sobey

978-1-55652-745-6
$14.95 (CAN $16.95)

"Perfect for collectors, for anyone daring enough to build homemade versions of these classic toys and even for casual browsers." –*Booklist*

Profiling 50 of the world's most popular playthings–including their history, trivia, and the technology involved–this guide uncovers the hidden science of toys. Discover how an Etch A Sketch writes on its gray screen, why a boomerang returns after it is thrown, and how an RC car responds to a remote control device. This entertaining and informative reference also features do-it-yourself experiments and tips on reverse engineering old toys to observe their interior mechanics, and even provides pointers on how to build your own toys using only recycled materials and a little ingenuity.

The Way Kitchens Work

The Science Behind the Microwave, Teflon Pan, Garbage Disposal, and More

Ed Sobey

978-1-56976-281-3
$14.95 (CAN $16.95)

If you've ever wondered how a microwave heats food, why aluminum foil is shiny on one side and dull on the other, or whether it is better to use cold or hot water in a garbage disposal, now you'll have your answers. *The Way Kitchens Work* explains the technology, history, and trivia behind 55 common appliances and utensils, with patent blueprints and photos of the "guts" of each device. You'll also learn interesting side stories, such as how the waffle iron played a role in the success of Nike, and why socialite Josephine Cochran *really* invented the dishwasher in 1885.

Available at your favorite bookstore, by calling (800) 888-4741, or at www.chicagoreviewpress.com